Elisha Kent Kane, Charles Anthony Schott

Tidal Observations in the Arctic Seas

Elisha Kent Kane, Charles Anthony Schott

Tidal Observations in the Arctic Seas

ISBN/EAN: 9783743314900

Manufactured in Europe, USA, Canada, Australia, Japa

Cover: Foto ©ninafisch / pixelio.de

Manufactured and distributed by brebook publishing software (www.brebook.com)

Elisha Kent Kane, Charles Anthony Schott

Tidal Observations in the Arctic Seas

Smithsonian Contributions to Knowledge.

TIDAL OBSERVATIONS

IN THE

ARCTIC SEAS.

BY

ELISHA KENT KANE, M.D., U.S.N.

MADE DURING THE SECOND GRINNELL EXPEDITION IN SEARCH OF SIR JOHN FRANKLIN,
IN 1853, 1854, AND 1855, AT VAN RENSSELAER HARBOR.

REDUCED AND DISCUSSED,

BY

CHARLES A. SCHOTT,
ASSISTANT U. S. COAST SURVEY.

WASHINGTON CITY:
PUBLISHED BY THE SMITHSONIAN INSTITUTION.
OCTOBER, 1860.
NEW YORK: D. APPLETON & CO.

CONTENTS.

	PAGE
INTRODUCTORY LETTER .	v
Explanatory and introductory remarks	1
Record of tidal observations at Van Rensselaer Harbor, 1853-4-5	5
Discussion of half-monthly inequality in time and height .	67
Effect of changes of the moon's declination and parallax .	72
Discussion of the diurnal inequality in height and time .	74
Investigation of the form of the tidal wave	78
Note on the effect of wind on the tides .	81
Note on the progress of the tidal wave and depth of the sea	81
Record of soundings . .	82
APPENDIX—containing a tidal record at Wolstenholm Sound, Commander Saunders, 1849-50, with four plates	83

INTRODUCTORY LETTER.

WASHINGTON, July 4th, 1860.

PROFESSOR JOSEPH HENRY, LL.D.,
    Secretary of the Smithsonian Institution:

DEAR SIR: The records of the tidal observations made under the direction of Dr. Kane, in the second Grinnell Expedition to the Arctic Regions, were placed in my hands by his late lamented father, Judge Kane, in December, 1857.

Dr. Kane had selected Assistant Charles A. Schott, of the U. S. Coast Survey, for the reduction of a considerable portion of the observations made on that expedition; and I, therefore, placed them in Mr. Schott's possession for reduction, and recommend his paper for publication in the "Smithsonian Contributions to Knowledge." It is proper to state that the computations were at the expense of the Smithsonian Institution. This is the sixth and last paper of the series.

Very respectfully, yours,
A. D. BACHE,
Superintendent U. S. Coast Survey.

RECORD AND REDUCTION OF THE TIDES.

The observations and discussion of the tides at Van Rensselaer Harbor, the winter quarters of the Advance during 1853-54 and 1854-55, will form the last of the series of papers on the results of the expedition, prepared by me for publication.

Occasional tidal observations were made after passing Smith Straits, when, owing to the peculiar navigation through the narrow openings between the coast and the bay ice, the vessel was much exposed to the tidal action, frequently grounding at low water, and otherwise, by taking advantage of high tides, slowly advancing to her winter quarters.

The bay, near the head of which the Advance was laid up, and used as the winter quarters by Dr. Kane's party, is freely exposed to the north (true) and northwest; the indentation of the shore line is about five miles; some rocky islands are situated within the bay.

Shortly after the vessel entered the harbor a tide staff was arranged, and a series of tidal observations was commenced on September 11, 1853, and continued, with occasional interruptions (partly owing to defects in the pulley-gauge, afterwards rigged up, and partly owing to other unavoidable accidents) till the 24th of January, 1855, on which date the regular log book appears to have been discontinued.

The several series of observations during this period are of very unequal value, as will appear in the detailed examination and discussion of the results. The difficulties to be overcome in the attempt to secure a reliable set of observations were considerable, those of a physical nature being the greatest. The observations with the staff or sounding line are subject to irregularities from a slow movement of the vessel, which, though imbedded in ice during the greater part of the year, is yet not stationary; these observations may also be affected by the softness of the bottom; the observations by means of a pulley tide gauge may be defective, on account of a slow drift of the vessel and motion of the ice field, also in consequence of a lengthening or shortening of the rope, or it may be in consequence of slipping of the rope on the circumference of the wheel. The latter defect, or one similar in its nature, has been a source of much annoyance, requiring the application of corrections to the readings, in order to refer all observations to the same zero of the scale. There is another defect to which pulley-gauges are subject, namely, the gradual rise of the vessel, in consequence of the consumption of provisions and fuel. Notices of these defects will appear in the subsequent discussion.

The pulley-gauge is described by Dr. Kane, in volume 1 of the *Narrative*, p. 117, as follows: "Our tide register was on board the vessel, a simple pulley-gauge, arranged with a wheel and index, and dependent on her rise and fall for its rotation."[1]

In order to ascertain the nature of the tides, as well as the degree of accuracy of the different observations, the readings were roughly plotted for a first examination; the following series were found suitable for discussion:—

SERIES I. *From October 10th, 1853, to December 28th, 1853.*—This series, with the exception of three days, is complete; the observations in the latter part of December appear to be of less reliable character. The observations between September 11 and October 4, 1853, are too fragmentary to be used. The pulley-gauge observations between October 4 and October 9 seem to have been only experimental. The hourly readings are superseded by half-hourly readings on November 8, and continue half hourly, day and night, to the end of the series. After November 28, corrective soundings were taken at noon each day. In order to make use of these soundings, the mean depth of the water at the anchorage was deduced from them as follows:—

	Mean reading.	
December, 1853.	43.8 feet, from 31 soundings (at noon).	
January, 1854.	44.9	21
February, "	44.3	17
March, "	43.3	19
April, "	41.8	20
May, "	43.5	9

The individual soundings will appear in the record following.

Mean depth of water at anchorage, in winter, 1853–54, 43.6 feet, as obtained from 117 soundings. The monthly mean values for the tidal level accord well, and show that no lateral change took place in the position of the brig (or else that the bottom was level). It will be seen that for Series I the reading 7.0 was adopted to express the mean level, the zero of the scale was, therefore, at an elevation of 36.6 feet from the bottom. The readings of the pulley gauge are expressed in feet,[2] as I have been informed by Mr. Sonntag.

SERIES II. *From January 28th, 1854, to April 7th, 1854.*—The double half-hourly readings of the pulley-gauge are continued. The series is complete with the exception of ten days, which had to be omitted. The register broke January 22d; observations commenced January 24th, but were not sufficiently regular for use

---

[1] The following note is appended: One end of the cord represented a fixed point, by being anchored to the bottom; the free end, with an attached weight, rose and fell with the brig, and recorded its motion on the grooved circumference of a wheel. This method was liable to objections, but it was corrected by daily soundings. The movements of our vessel partook of those of the floe in which she was imbedded, and were unaccompanied by any lateral deviation.

[2] The following is an extract from Mr. Sonntag's letter to me, dated New York, March 23, 1860: "The circumference of the wheel (of the pulley-gauge) was divided into feet and tenths of a foot, and the records by the sounding line are also expressed in feet and decimals. The records of the wheel are very uncertain, as often the rope slid over the wheel without turning it, owing to the ice which surrounded the axis."

until January 28th. The corrective soundings at noon are continued, with occasional omissions, throughout this series. After April 7th there is a break in the observations, those between the 14th and 20th appear to be irregular.

SERIES III. *From April 20th, 1854, to August 3d, 1854.*—The double half-hourly readings of the pulley-gauge continue to May 5th, after which date single half-hourly readings are recorded. The corrective soundings cease on the 12th of May. Interruptions occur between May 4th and May 7th, also on July 8th, also between July 15th and 18th, and between July 20th and the 28th. On the 8th of August the brig was released from her ice cradle, and rose two and a half feet; occasional warpings of the brig after this date render the observations worthless. On the 23d of August the brig was in but seven feet of water, and grounded.

SERIES IV. *From September 7th, 1854, to October 22d, 1854.*—The hourly observations assume again a more regular appearance on the 7th of September; they were taken with the sounding line, and are expressed in fathoms and feet (as stated in a note, August 12th). The following note is of October 21st, 1854: "The tide register as yet not rigged, observations very faulty by sounding line." The irregularities increase after this date; on the 15th of November following, the tide register was arranged, and observations (hourly) commenced on the 17th; the slipping of the rope, however, was of so frequent occurrence and of so great an extent, that it was considered better to take no further notice of these observations; the record continues to January 24th, 1855, when the strength of the party no longer permitted due attention to the tidal phenomena.

It was apparent that before any closer insight into the nature of these tides could be obtained, they must first be reduced to the same zero or mean level of the sea. To effect this in a manner apparently best suiting the case, and otherwise unobjectionable, two curved lines were traced on the diagrams, the upper one enveloping the highest high water of each day, the other enveloping the lowest low water of each day; in tracing these lines some allowance was made, when necessary, for disturbing causes, so as to obtain tolerably smooth curves; cases of abrupt changes were, of course, treated accordingly. A line, equidistant from these curves, was assumed as representing the mean level, and when straightened out was adopted as axis of the mean level of the sea. The corrections to refer each observation to this adopted mean level; or, in other words, the corrections required to refer each observation to the same zero of the scale, so as to make them comparable with each other, were taken from the projection, and are given in the column headed "reduction," in the following record.

This method of treatment excludes necessarily in Series I, II, and III, any discussion of the variation in the mean level of the sea, the oscillations of which have been found small at other places. As an illustration of this, the tides at Singapore might be referred to; the Rev. W. Whewell (7th series of researches on the tides, *Phil. Trans. of the Roy. Soc.*, Part I, 1837), finds for these tides that, if a line is drawn representing the mean height (midway between high and low water each day) it is very nearly constant, though the successive low waters often differ by six feet (on account of the diurnal inequality), the mean level only oscillates through a few inches. It appears from Mr. Lloyd's paper (*Phil. Trans.* of 1831) that the mean level at Sheerness is higher in spring tides than in neap tides by seven inches nearly; also there seems to be no doubt (as shown by Mr. Whewell, *Phil. Trans.*, 1839 and 1840) that the mean level increases as the moon's declination increases, amounting to three inches at Plymouth, when the moon's declination is 25°; at Petropaulofsk and Novo-Arkhangelsk the mean level rises as the moon's declination increases.

The use of the soundings intended to furnish corrections to the readings of the pulley-gauge is in many cases a doubtful remedy, on account of the continued change in the zero of the wheel's index; in fact, it would have required numerous soundings at other hours than noon. As it is, a combination of the corrections by enveloping curves and soundings had to be adopted. Thus, for December 5th, soundings at noon 43.0 feet (see record further on), mean level 36.6, hence reading of scale at noon 6.4; reading of pulley-gauge at that hour 19.0, correction by curve —12.5, corrected reading 6.5, which agrees with the first number; this is, however, a very favorable case. For intermediate hours the correction as given by the curves serve as guides. The reduction to the same level affects the times generally very little.

The following table contains the soundings taken at noon between the interval of the first and second series, those taken during the series being given in the record.

SOUNDINGS AT NOON.

1853.	Fath.	Feet.	Inch.	Register.	1854.	Fath.	Feet.	Inch.	Register.
December, 29.	7	3	0		January 13.	7	3	6	
30.	8	0	0	18.1 (changed.)	14.	...	...	...	
31.	8	2	0		15.	8	1	0	
1854. Jan. 1.	8	1	6		16.	7	2	6	
2.	8	1	6		17.	...	...	...	
3.	7	5	6		18.	7	3	9	
4.	7	3	0	Changed to 16.0	19.	7	5	6	
5.	7	1	6		20.	6	3	0	
6.	6	4	6	Changed to 10.5	21.	6	4	0	Changed to 10
7.	6	3	0		22.	Tide register broken.			
8.	...	...	...		23.	"	"	"	
9.	6	4	2		24.	"	"	"	
10.	7	0	0		25.	...	...	...	
11.	...	...	...		26.	...	...	...	
12.	7	4	0		27.	7	1	9	

The following soundings were taken between the second and third series:—

1854.	Fath.	Feet.	Inches.	1854.	Fath.	Feet.	Inches.
April 8.	6	5	6	April 16.	7	5	6
9.	6	4	0	(Fall 15 feet 8 inches.)			
10.	7	0	6	(17.	6	5	0
11.	6	5	6	at 20 minutes to 5.)			
13.	7	4	0	(18.	6	0	0
14.[1]	7	5	6	at 8^h 15^m P. M.)			
15.	8	0	0	19.	6	2	0

(Low water to high water 14 ft. 8 inch.)

---

[1] For the past ten days the tide register has not been reliable on account of the rope slipping.

The note of February 3d, 1854, is very instructive in regard to the effect of the tides on the ice floe, viz: "The enormous elevation of the land ice by the tides has raised a barrier of broken tables seventy-two feet wide and twenty feet high between the brig and islands. This action has caused a recession of the main floe; our vessel has changed her position twenty feet within the last two spring tides, and the hawser connected with Butler Island parted with the strain." The cutwater of the brig was then 280 feet from the margin of the ice. (Note of February 4th.)

The mean of all the soundings taken during the fourth series is very nearly fifteen feet, hence the constant index error, to refer the observations to the level previously adopted, is eight feet, which correction was applied, converting at the same time the record of fathoms into feet.

The following tidal record extends, therefore, over about nine and a half lunations between October 10, 1853, and October 22, 1854, during which interval the time and height of nearly five hundred high and as many low waters were secured.

Record of the Observations of the Tides at Van Rensselaer Harbor, North Greenland, in 1853, 1854, and 1855.

POSITION OF THE WINTER QUARTERS,
Latitude 78° 37' north, and longitude 70° 53', or 4^h $43^m.5$ west of Greenwich.[1]

The first column for each day is copied from the original log-book, the second column contains the reduction to the adopted zero of scale found graphically as explained, and the third column contains the observations referred to the same mean level.

---

[1] See my discussion of the astronomical observations of the expedition in vol. XII of the Smithsonian Contributions to Knowledge, 1860.

RECORD AND REDUCTION OF THE TIDES.

SERIES 1.—TIDAL OBSERVATIONS FROM OCTOBER 10, 1853, TO DECEMBER 28, 1853.
Hourly observations on the pulley-gauge. Adopted reading of mean level 7.0, expressed in units of the scale. Increasing numbers indicate rise of water.

October, 1853.

Mean solar hour.	10th.	Red. to level.	Ref. obs.	11th.	Red. to level.	Ref. obs.	12th.	Red. to level.	Ref. obs.	13th.	Red. to level.	Ref. obs.	14th.	Red. to level.	Ref. obs.	17th.	Red. to level.	Ref. obs.
1	...	...	...	5.4	—1.0	4.4	5.0	—1.0	4.0	7.5	—1.3	6.2	7.0	—1.5	5.5	...		...
2	6.6	—1.0	5.6	5.4	"	4.4	5.0	"	4.0	6.4	"	5.1	5.4	"	3.9	...		...
3	7.6	"	6.6	5.5	"	4.5	5.2	"	4.2	5.0	"	3.7	4.4	"	2.9	...		...
4	8.1	"	7.1	6.0	"	5.0	6.0	"	5.0	5.5	"	4.2	4.3	"	2.7	...		...
5	8.7	"	7.7	6.8	"	5.9	7.0	"	6.0	7.9	"	6.6	4.4	—1.6	2.8	...		...
6	8.7	"	7.7	7.3	"	6.3	8.5	"	7.5	8.3	"	7.0	5.5	"	3.9	...		...
7	8.7	"	7.7	7.7	"	6.7	8.9	"	7.9	8.7	—1.4	7.3	7.5	"	5.9	...		...
8	9.0	"	8.0	7.7	"	6.7	8.9	"	7.9	9.4	"	8.0	9.6	"	8.0			
9	6.4	"	5.4	7.6	"	6.6	...		...	10.5	"	9.1	11.2	"	9.6			
10	5.9	"	4.9	6.6	"	5.6	7.8	—1.1	6.7	10.5	"	9.1	11.4	"	9.8			
11	5.7	"	4.7	6.1	"	5.1	6.7	"	5.6	10.3	"	8.9	11.3	"	9.7			
Noon	5.8	"	4.8	5.8	"	4.8	5.6	"	4.5	9.9	"	8.5	11.0	—1.7	9.3			
1	6.7	"	5.7	5.8	"	4.8	5.3	"	4.2	7.6	"	6.2	9.4	"	7.7			
2	7.3	"	6.3	5.8	"	4.8	5.3	"	4.2	6.7	"	5.3	7.4	"	5.7			
3	8.9	"	7.9	6.3	"	5.3	5.3	"	4.2	5.6	—1.5	4.1	6.6	"	4.9			
4	9.3	"	8.3	7.7	"	6.7	5.3	"	4.2	4.6	"	3.1	4.4	"	2.7	...		...
5	10.2	"	9.2	9.0	"	8.0	6.4	—1.2	5.2	6.5	"	5.0	4.6	"	2.9	4.2	—2.7	1.5
6	10.2	"	9.2	10.1	"	9.1	7.8	"	6.6	9.0	"	7.5	5.0	"	4.2	4.5	"	1.8
7	10.2	"	9.2	10.5	"	9.5	9.0	"	8.7	10.5	"	9.0	9.2	"	7.5			
8	9.9	"	8.9	10.5	"	9.5	11.0	"	9.8	11.6	"	10.1	12.0	—1.8	10.2	9.5	"	6.8
9	8.7	"	7.8	9.8	"	8.8	11.3	—1.3	10.0	12.4	"	10.9	12.4	"	10.6	12.6	"	9.9
10	7.5	"	6.5	9.0	"	8.0	11.3	"	10.0	12.4	"	10.9	13.1	"	11.3	13.0	"	10.3
11	6.3	"	5.3	7.2	"	6.2	9.7	"	8.4	12.4	"	10.9	13.1	"	11.3	13.4	"	10.7
Midn't	5.7	"	4.7	5.6	"	4.6	8.3	"	7.0	10.4	"	8.9	13.0	—1.9	11.1	13.4	"	10.7

October, 1853.

Mean solar hour.	18th.	Red. to level.	Ref. obs.	19th.	Red. to level.	Ref. obs.	20th.	Red. to level.	Ref. obs.	21st.	Red. to level.	Ref. obs.	22d.	Red. to level.	Ref. obs.	23d.	Red. to level.	Ref. obs.
1	11.5	—2.7	8.8	13.6	—2.2	11.4	13.8	—2.3	11.5	11.6	—1.7	9.9	10.7	—0.2	10.5	7.0	+0.8	7.8
2	8.9	"	6.2	12.0	"	9.8	12.6	"	10.3	"		...	10.2	"	10.0	8.0	"	8.8
3	6.8	"	4.1	8.9	"	6.7	10.5	"	8.2	10.8	—1.6	9.2	10.0	—0.1	9.9	9.5	"	10.3
4	5.5	"	2.8	6.6	"	4.4	8.0	"	5.7	9.6	—1.5	8.1	9.0	"	8.9	9.0	+0.9	9.9
5	4.4	—2.6	1.8	4.5	"	2.3	7.9	—2.2	5.7	6.7	—1.4	5.3	8.0	0.0	8.0	6.9	"	7.8
6	4.4	"	1.8	3.8	"	1.6	7.9	"	5.7	4.6	"	3.2	7.0	"	7.0	4.5	"	5.4
7	6.5	"	3.9	3.8	"	1.6	7.9	"	5.7	4.1	—1.3	2.8	5.0	"	5.0	3.4	"	4.3
8	8.7	"	6.1	4.7	"	2.5	8.9	"	6.7	4.1	"	2.9	5.0	+0.1	5.1	3.0	"	3.9
9	11.8	—2.5	9.3	5.3	"	3.1	8.7	"	6.5	6.7	"	5.5	3.3	"	3.4	3.1	"	4.0
10	...		...	11.0	"	8.8	8.7	"	6.5	8.5	—1.1	7.4	4.5	+0.2	4.7	3.5	"	4.4
11	...		...	13.6	"	11.4	12.9	"	10.7	10.8	"	9.7	6.3	"	6.5	3.7	"	4.6
Noon	...		...	14.4	"	12.2	13.6	"	11.4	11.7	—1.0	10.7	7.2	"	7.4	4.5	"	5.4
1	...		...	14.6	"	12.4	14.0	—2.1	11.9	11.9	"	10.9	7.0	+0.3	7.3	6.5	"	7.4
2	14.8	"	12.3	12.6	"	10.4	14.0	"	11.9	11.9	—0.9	11.0	7.5	"	7.8	7.0	"	7.2
3	10.6	—2.4	8.2	10.4	"	8.2	...		...	11.9	—0.8	11.1	9.5	+0.4	9.9	8.5	"	9.4
4	8.6	"	6.2	9.6	—2.3	7.3	...		...	11.8	"	11.0	9.2	"	9.6	9.5	"	10.4
5	6.6	"	4.2	6.6	"	4.3	7.7	"	5.6	9.0	—0.7	8.3	9.2	"	9.6	8.0	+0.8	8.8
6	4.4	"	2.0	5.2	"	2.9	6.2	"	4.1	7.5	"	6.8	6.7	+0.5	7.2	7.4	"	8.2
7	4.4	—2.3	2.1	4.2	"	1.9	5.5	"	3.4	5.5	—0.6	4.9	5.2	"	5.7	7.0	"	7.8
8	5.5	"	3.2	4.8	"	2.5	5.2	—2.0	3.2	5.0	"	4.4	3.9	+0.6	4.5	6.1	"	6.9
9	5.3	"	6.0	6.8	"	4.5	4.7	"	2.7	5.0	—0.5	4.5	3.9	"	4.5	5.5	"	6.3
10	10.4	"	8.1	9.4	"	7.1	6.8	—1.9	4.9	5.0	"	4.5	4.0	"	4.6	4.5	+0.7	5.2
11	12.6	"	10.3	11.6	"	9.3	9.8	"	7.9	5.0	—0.4	4.6	6.3	+0.7	7.0	5.0	"	5.7
Midn't	13.7	—2.2	11.5	13.4	"	11.1	11.0	—1.8	9.2	5.0	—0.3	4.7	...		...	5.5	"	6.2

Regular observations commence October 10, 2 A. M.
Oct. 15. Tide rope found broken at 10 A. M., and the lead lost through the ice hole.
Oct. 17. Tides irregular, index changed 12 units; hence most of the observations on this day had to be omitted.
Oct. 20. The observation for 10 A. M. is incorrect, on account of obstruction by the ice.
Oct. 21. Flood (rise) commenced at 8 P. M.
Oct. 24. Slack water (stand) at 8 o'clock, flood commences.

RECORD AND REDUCTION OF THE TIDES.

Series I.—Tidal Observations from October 10, 1853, to December 28, 1853.
Hourly observations on the pulley-gauge. Adopted reading of mean level 7.0, expressed in units of the scale. Increasing numbers indicate rise of water.

October, 1853.

Mean solar hour.	24th.	Red. to level.	Ref. obs.	25th.	Red. to level.	Ref. obs.	26th.	Red. to level.	Ref. obs.	27th.	Red. to level.	Ref. obs.	28th.	Red. to level.	Ref. obs.	29th.	Red. to level.	Ref. obs.
1	6.5	+0.7	7.2	5.5	−0.1	5.4	5.0	+0.3	5.3	4.0	+0.8	4.8	4.5	+1.1	5.6	5.1	+0.9	6.0
2	7.0	"	7.7	5.8	"	5.7	5.5	"	5.8	4.0	"	4.8	3.5	"	4.6	4.4	"	5.3
3	8.0	"	8.7	6.5	−0.2	6.3	5.5	+0.4	5.9	4.5	"	5.3	3.4	"	4.5	2.8	"	3.7
4	8.0	+0.6	8.6	7.2	"	7.0	5.8	"	6.2	5.0	+0.9	5.9	4.3	"	5.4	2.5	"	3.4
5	7.3	"	7.9	7.3	"	7.1	5.8	"	6.2	6.0	"	6.9	5.3	"	6.4	3.0	"	3.9
6	6.5	"	7.1	7.0	−0.3	6.7	6.2	+0.5	6.7	6.8	"	7.7	5.6	"	6.7	5.5	"	6.4
7	5.4	"	6.0	6.4	"	6.1	6.5	"	7.0	7.1	+1.0	8.1	7.3	"	8.4	7.0	"	7.9
8	4.0	+0.5	4.5	5.6	"	5.3	6.5	"	7.0	7.1	"	8.1	8.2	"	9.3	7.8	"	8.7
9	5.0	"	5.5	5.6	−0.4	5.2	6.1	"	6.6	7.1	"	8.1	---	---	---	9.8	"	10.7
10	5.0	"	5.5	5.6	"	5.2	5.6	"	6.1	6.0	"	7.0	---	---	---	9.8	"	10.7
11	5.0	"	5.5	5.6	"	5.2	5.5	+0.6	5.6	5.6	"	6.6	---	---	---	9.6	"	10.5
Noon	5.0	+0.4	5.4	5.6	"	5.2	5.5	"	6.1	4.5	"	5.5	---	---	---	9.3	"	10.2
1	7.7	"	8.1	7.3	−0.3	7.0	5.5	"	6.1	4.5	"	5.5	4.8	"	5.9	6.8	"	7.7
2	---	---	---	6.3	"	6.0	5.5	"	6.1	5.0	"	6.0	4.3	"	5.4	4.8	"	5.7
3	---	---	---	6.8	−0.2	6.6	6.0	"	6.6	5.3	"	6.3	4.0	"	5.1	4.0	"	4.9
4	---	---	---	7.5	"	7.3	6.5	"	7.1	6.3	"	7.3	4.0	"	5.1	4.0	"	4.9
5	9.5	+0.2	9.7	7.3	−0.1	7.2	7.5	+0.7	8.2	7.7	"	8.7	6.0	"	7.1	4.0	"	4.9
6	9.5	"	9.7	7.6	"	7.5	8.0	"	8.7	7.3	+1.1	8.4	7.1	"	8.2	5.5	"	6.4
7	8.1	+0.1	8.2	8.3	"	8.2	8.2	"	8.9	8.6	"	9.7	8.0	+1.0	9.0	6.7	"	7.6
8	7.0	"	7.1	8.5	0.0	8.5	8.5	"	9.2	8.7	"	9.8	9.4	"	10.4	9.0	"	9.9
9	6.0	0.0	6.0	7.1	"	7.1	8.5	"	9.2	8.7	"	9.8	9.8	"	10.8	9.5	"	10.4
10	5.8	"	5.8	6.1	+0.1	6.2	7.1	"	7.8	7.1	+1.2	8.3	9.8	"	10.8	10.5	"	11.4
11	5.5	−0.1	5.4	5.4	"	5.5	6.1	+0.8	6.9	6.3	"	7.5	9.0	+0.9	9.9	10.5	"	11.4
Midn't	5.5	"	5.4	5.0	+0.2	5.2	4.8	"	5.6	5.9	"	7.1	7.3	"	8.2	8.5	"	9.4

October, 1853. /// November, 1853.

Mean solar hour.	30th.	Red. to level.	Ref. obs.	31st.	Red. to level.	Ref. obs.	1st.	Red. to level.	Ref. obs.	2d.	Red. to level.	Ref. obs.	3d.	Red. to level.	Ref. obs.	4th.	Red. to level.	Ref. obs.
1	4.5	+0.9	5.4	5.4	+1.5	6.9	6.8	+1.7	8.5	9.1	+1.4	10.5	13.5	−1.7	11.8	13.0	−1.4	11.6
2	3.5	"	4.4	3.8	"	5.3	4.0	"	5.7	8.8	+1.3	10.1	12.0	"	10.3	12.5	"	11.1
3	2.0	"	2.9	3.0	+1.6	4.6	2.5	"	4.2	7.1	+1.2	8.3	8.0	"	7.3	11.2	"	9.8
4	2.0	"	2.9	0.0	"	1.6	1.0	"	2.7	4.7	+1.0	5.7	6.7	−1.6	5.1	9.8	"	8.4
5	3.2	"	4.1	1.0	"	2.6	1.0	"	2.7	1.5	+0.7	2.2	3.7	"	2.1	6.0	"	4.6
6	4.2	"	5.1	1.2	"	2.8	2.2	"	3.9	0.5	+0.5	1.0	1.8	"	0.2	4.4	"	3.0
7	6.2	+1.0	7.2	3.1	+1.7	4.8	3.2	"	4.9	0.7	+0.2	0.9	1.5	−1.5	0.0	2.0	"	0.6
8	8.5	"	9.5	6.5	"	8.2	7.0	"	8.7	3.0	0.0	3.0	2.4	"	0.9	1.7	"	0.3
9	10.1	+1.1	11.2	8.5	"	10.2	9.5	+1.6	11.1	7.2	−0.2	7.0	3.4	"	1.9	3.7	"	2.3
10	10.4	"	11.5	9.3	"	11.0	10.5	"	12.1	9.8	−0.5	9.3	4.2	"	2.7	7.3	"	5.9
11	10.4	"	11.5	10.3	"	12.0	10.7	"	12.3	10.9	−0.7	10.2	6.0	"	4.5	10.5	"	9.1
Noon	10.4	+1.2	11.6	10.0	"	11.7	10.7	"	12.3	15.3	−1.0	14.3	6.7	"	5.2	12.6	"	11.2
1	8.2	"	9.4	7.5	"	9.2	10.2	"	11.8	15.2	−1.2	14.0	15.7	"	14.2	---	---	---
2	5.5	"	6.7	5.2	"	6.9	9.2	"	10.8	13.6	−1.5	12.1	15.2	"	13.7	---	---	---
3	3.7	"	4.9	3.7	"	5.4	8.1	"	9.7	10.5	−1.6	8.9	12.5	"	11.0	---	---	---
4	3.4	"	4.6	0.1	"	1.8	4.5	+1.5	6.0	6.8	−1.7	5.1	10.1	"	8.6	12.0	"	10.6
5	2.6	+1.3	3.9	0.2	"	1.9	3.0	"	4.5	3.8	"	2.1	6.5	"	5.0	11.5	"	10.1
6	3.4	"	4.7	1.2	"	2.9	2.3	"	3.8	2.0	"	0.3	4.5	"	3.0	9.0	"	7.6
7	5.1	"	6.4	3.0	"	4.7	2.1	"	3.6	1.8	"	0.1	3.5	"	2.0	5.5	"	4.1
8	6.8	"	8.1	5.2	"	6.9	4.0	"	5.5	3.2	"	1.5	3.5	"	2.0	3.0	"	1.6
													3.0	"	1.5	3.0	"	1.6
9	9.5	+1.4	10.9	8.1	"	9.8	9.0	"	10.5	3.3	"	1.6	4.5	"	3.0	3.1	"	1.7
10	10.3	"	11.7	9.5	"	11.2	10.1	"	11.6	3.3	"	1.6	7.3	"	5.8	5.0	"	3.6
11	10.3	"	11.7	9.5	"	11.2	10.1	"	11.6	3.3	"	1.6	9.3	"	7.8	7.5	"	6.1
Midn't	9.0	+1.5	10.5	10.5	"	12.2	10.1	"	11.6	3.3	"	1.6	12.0	"	10.5	9.8	"	8.4

Oct. 29. Slack water [stand] of ebb at 4h 30m A. M.
Oct. 31. Slack water [stand] of ebb at 5 A. M.
Nov. 2 to Nov. 6. Between these dates there are occasionally half-hourly readings, but unless they occur near high or low water they are omitted in the above.

RECORD AND REDUCTION OF THE TIDES.

SERIES I.—TIDAL OBSERVATIONS FROM OCTOBER 10, 1853, TO DECEMBER 28, 1853.

Hourly observations on the pulley-gauge. Adopted reading of mean level 7.0, expressed in units of the scale. Increasing numbers indicate rise of water.

November, 1853.

Mean solar hour.	5th.	Red. to level.	Ref. obs.	6th.	Red. to level.	Ref. obs.	7th.	Red. to level.	Ref. obs.	8th.	Red. to level.	Ref. obs.	9th.	Red. to level.	Ref. obs.	10th.	Red. to level.	Ref. obs.
1	11.5	—1.4	10.1	9.0	—1.0	8.0	6.6	—0.6	6.0	4.3 5.3 5.6	—0.5	3.8 4.8 5.1	3.8 3.9 4.2	+0.2 " "	4.0 4.1 4.4	4.5 4.5 4.3	—0.2 " "	4.3 4.3 4.1
2	11.0	"	9.6	9.8 9.9 8.8	" " "	8.8 8.9 8.8	7.9	" "	7.3 8.2	6.2	"	5.7	4.0 4.5	" "	4.2 4.7	4.0 4.2	" "	3.8 4.0
3	10.0	"	8.6	10.0 10.0	" "	9.0 9.0	8.8 8.8	" "	8.2 8.2	7.1	"	6.6	5.0	"	5.2	4.5 4.5	—0.3	4.2 4.2
4	9.2	—1.3	7.9	10.0 8.1	—0.9 "	9.1 7.2	8.8 8.6	" "	8.2 8.2	7.9 7.8	" —0.4	7.4 7.5	6.0	"	6.2	4.1 5.8	" "	3.8 5.5
5	5.7	"	4.4	7.7	"	6.8	8.6	"	8.0	8.2 8.0	" "	7.8 7.6	7.5	"	7.7	6.5	"	6.2
6	3.5	"	2.2	5.5	"	4.6	7.8	—0.5	7.3	7.8	"	7.4	7.9 7.9	" "	8.1 8.1	8.5 9.9	" "	8.2 9.6
7	2.3	"	1.0	4.5 4.0	" "	3.6 3.1	6.5	"	6.0	7.3	—0.3	7.0	8.0 8.0	+0.1 "	8.1 8.1	9.5 10.0	" "	9.2 9.7
8	2.0 2.0	" "	0.7 0.7	3.6 3.1	—0.8 "	2.8 2.3	5.4	"	4.9	6.3	"	6.0	8.0 7.4	" "	8.1 7.5	10.2 10.2	—0.4 "	9.8 9.8
9	2.4 2.6	" "	1.1 1.3	3.1 3.2	" "	2.3 2.4	4.0 4.0	" "	3.5 3.5	5.5	"	5.2	6.9	"	7.0	10.2 10.1	" "	9.8 9.7
10	2.7	—1.2	1.5	4.3	"	3.5	3.7 3.4	" "	3.2 3.4	5.3	"	5.0	6.0	"	6.1	9.8	"	9.4
11	6.7	"	5.5	5.7	—0.7	5.0	4.4 5.5	" "	3.9 5.0	5.5 5.3	" —0.2	5.2 5.1	4.8	"	4.0	9.0	"	8.6
Noon	11.1	"	9.9	8.3	"	7.6	6.8	"	6.3	5.1 5.3	" "	4.9 5.1	4.6	0.0	4.6	7.5	—0.5	7.0
1	13.1 13.6	" "	11.9 12.4	...			8.1	"	7.6	5.6	"	5.4	4.4 4.5	" "	4.4 4.5	6.5 6.3	" "	6.0 5.8
2	14.2 14.2	" "	13.0 13.0	...			9.5	"	9.0	7.1	"	6.9	4.5 5.1	" "	4.5 5.1	6.0 6.0	" "	5.5 5.5
3	14.1 13.0	—1.1 "	13.0 11.9	...			10.6 10.9	" "	10.1 10.4	8.3	"	8.1	5.5	"	5.5	6.5	—0.6	5.9
4	12.3	"	11.2	11.5 11.0	—0.6 "	10.9 10.4	11.2 11.1	" "	10.7 10.6	9.5	"	9.3	7.1	"	7.1	7.0	"	6.4
5	10.8		9.7	10.7	"	10.1	10.2	"	9.7	10.1 10.5	—0.1 "	10.0 10.4	9.2 9.6	—0.1	9.1 9.5	7.9	"	7.3
6	8.2	"	7.1	9.1	"	8.5	10.2	"	9.7	10.5 10.5	" "	10.4 10.4	10.1 9.5	"	10.0 9.4	9.5	"	8.9
7	5.1	"	4.0	7.5	"	6.9	8.5	"	8.0	10.5 10.5	" "	10.4 10.4	11.1 11.1	"	11.0 11.0	10.9 11.4	" "	10.3 10.8
8	3.3 2.9	" "	2.2 1.8	5.4	"	4.8	7.5	"	7.0	10.0	0.0	10.0	11.0 9.8	"	10.9 9.7	11.0 11.0	" "	10.4 10.4
9	3.0 3.5	" "	1.9 2.4	4.2 4.0	" "	3.6 3.4	5.6	"	5.1	8.1	"	8.1	8.9	"	8.8	11.0 11.3	" "	10.4 10.7
10	4.1	—1.0	3.1	3.2 3.2	" "	2.6 2.6	4.4 4.3	" "	3.9 3.8	7.0	+0.1	7.1	7.0	"	6.9	11.3 10.9	—0.7 "	10.6 10.2
11	6.0	"	5.0	4.0 4.5	" "	3.4 3.9	4.1 4.1	" "	3.6 3.6	4.9 4.2	" "	5.0 4.3	4.7 4.7	—0.2 "	4.5 4.5	9.5	"	8.8
Midn't	7.3	"	6.3	4.9	"	4.3	4.0	"	3.5	4.0	+0.2	4.2	4.5	"	4.3	8.5	"	7.8

Nov. 8. From this date the observations are half-hourly : in the above record, however, only those half-hourly readings were inserted, which occur near a high or low water.

RECORD AND REDUCTION OF THE TIDES.

SERIES I.—TIDAL OBSERVATIONS FROM OCTOBER 10, 1853, TO DECEMBER 28, 1853.

Hourly observations on the pulley-gauge. Adopted reading of mean level 7.0, expressed in units of the scale. Increasing numbers indicate rise of water.

November, 1853.

Mean solar hour.	11th.	Red. to level.	Ref. obs.	12th.	Red. to level.	Ref. obs.	13th.	Red. to level.	Ref. obs.	14th.	Red. to level.	Ref. obs.	15th.	Red. to level.	Ref. obs.	16th.	Red. to level.	Ref. obs.
	6.1	—0.7	5.4													10.8	—5.5	11.3
1	4.6	"	3.9	6.0	—0.9	5.1	8.8	—1.6	7.2	11.5	—4.1	7.4	13.0	—5.2	7.8	15.1	"	9.6
	4.4	"	3.7															
2	3.6	—0.8	2.8	5.5	"	4.6	6.7	"	5.1	9.1	—4.2	4.9	10.0	"	4.8	14.7	—5.6	9.1
	3.1	"	2.3	4.5	"	3.6												
3	3.5	"	2.7	4.0	—1.0	3.0	4.9	—1.7	3.2	7.5	—4.3	3.2	8.0	"	2.8	13.0	"	7.4
	3.6	"	2.8	4.0	"	3.0	4.0	"	2.9	7.0	"	2.7						
4	4.1	"	3.3	3.8	"	2.6	4.4	—1.8	2.6	6.5	—4.4	2.1	6.6	"	1.4	10.5	"	4.0
				4.1	"	3.1	4.5	"	2.7	6.5	"	2.1	6.0	"	0.8	6.1	—5.7	0.4
5	5.0	"	4.2	4.6	"	3.6	4.5	—1.9	2.6	6.5	"	2.1	5.9	"	0.7	6.1	"	0.4
							5.0	"	3.1	6.5	"	2.1	5.9	"	0.7	6.3	—5.8	0.5
6	5.6	"	4.8	7.1	"	6.1	5.5	"	3.6	7.0	—4.5	2.5	6.2	"	1.0	6.3	"	0.5
7	8.0	"	7.2	8.6	"	7.6	7.5	—2.0	5.5	9.0	"	4.5	8.0	"	2.8	7.3	—5.9	1.4
	8.3	"	7.5															
8	9.5	"	8.7	10.5	"	9.5	10.2	—2.1	8.1	12.4	—4.6	7.8	10.5	"	5.3	10.0	—6.0	4.0
	10.3	"	9.5	11.5	"	10.5												
9	10.3	"	9.5	12.0	"	11.0	12.7	"	10.6	15.5	"	10.9	12.9	"	7.7	12.7	"	6.7
	10.0	"	9.2	12.2	"	11.2	13.0	—2.2	10.8									
10	10.0	"	9.2	12.2	"	11.2	13.7	"	11.5	17.0	—4.7	12.3	16.0	"	10.8	16.4	—6.1	10.3
				12.0	"	11.0	13.7	"	11.4	17.5	—4.8	12.7	17.3	"	12.1			
11	9.5	"	8.7	11.7	—1.1	10.6	13.5	"	11.2	17.3	—4.9	12.4	18.5	"	13.3	17.9	"	11.8
							12.9	"	10.6	17.1	"	12.2	16.5	"	13.3	18.5	—6.2	12.3
Noon	7.5	"	6.7	9.0	"	7.9	12.1	—2.4	9.7	16.4	—5.0	11.4	18.3	—5.3	13.0	19.9	—6.3	13.6
													18.0	"	12.7	19.9	"	13.6
1	6.0	"	5.2	8.0	"	6.9	11.1	—2.5	8.6	14.1	"	9.1	17.6	"	12.3	18.7	—6.4	12.3
2	5.3	"	4.5	6.3	"	5.2	9.1	—2.6	6.5	12.0	"	7.0	15.5	"	10.2	17.9	"	11.5
	5.3	"	4.5	5.6	"	4.5												
3	5.0	"	4.2	4.9	"	3.8	7.8	—2.7	5.1	9.3	—5.1	4.2	12.0	"	6.7	15.0	"	8.6
	5.0	"	4.2	5.0	"	3.9												
4	5.5	"	4.7	5.0	"	3.9	6.3	—2.8	3.5	8.5	"	3.4	10.3	"	5.0	12.6	—6.5	6.1
				5.0	"	3.9				9.0	"	3.9	9.5	"	4.2			
5	6.6	"	5.5	5.5	—1.2	4.3	6.3	—2.9	3.4	8.0	"	2.9	8.9	"	3.6	10.6	"	4.1
							5.5	"	2.6	8.0	"	2.9	8.9	"	3.6	10.3	—6.7	3.6
6	8.0	"	7.2	6.4	"	5.2	5.5	—3.0	2.5	8.5	—5.2	3.3	8.9	—5.4	3.5	10.0	"	3.3
							6.8	—3.1	3.7				8.9	"	3.5	10.0	—6.8	3.2
7	10.0	"	9.2	7.7	"	6.5	8.7	—3.2	5.5	9.6	"	4.4	9.2	"	3.8	10.0	—6.9	3.1
																11.0	"	4.1
8	11.4	—0.9	10.5	10.4	—1.3	9.1	11.0	—3.3	7.7	11.4	"	6.2	10.7	"	5.3	11.5	—7.0	4.5
	11.8	"	10.9															
9	12.7	"	11.8	11.8	"	10.5	13.7	—3.4	10.3	13.1	"	7.9	13.7	"	8.3	14.0	"	7.0
	12.7	"	11.5	12.8	"	11.5												
10	12.1	"	11.2	12.7	—1.4	11.3	14.4	—3.6	10.8	14.7	"	9.5	16.6	"	11.2	16.0	"	9.0
				12.5	"	11.1	14.5	—3.7	10.8	15.0	"	9.8						
11	11.4	"	10.5	11.9	"	10.5	14.8	—3.8	11.0	14.6	"	9.4	17.8	"	12.4	17.5	—7.1	10.4
							14.4	—3.9	10.5	14.5	"	9.3	18.5	—5.5	13.0	18.0	"	10.9
Midn't	9.2	"	8.3	10.6	—1.5	9.1	13.0	—4.0	9.6	14.1	"	8.9	18.5	"	13.0	17.9	"	10.8

RECORD AND REDUCTION OF THE TIDES.

Series I.—Tidal Observations from October 10, 1853, to December 28, 1853.

Hourly observations on the pulley-gauge. Adopted reading of mean level 7.0, expressed in units of the scale. Increasing numbers indicate rise of water.

November, 1853.

Mean solar hour.	17th.	Red. to level.	Ref. obs.	18th.	Red. to level.	Ref. obs.	19th.	Red. to level.	Ref. obs.	20th.	Red. to level.	Ref. obs.	21st.	Red. to level.	Ref. obs.	22d.	Red. to level.	Ref. obs.
1	17.8 / 17.0	−7.2 "	10.6 / 9.8	1.4 / 1.0 / 20.3	+12.3 " / −7.6	13.7 / 13.3 / 12.7	16.5 / 16.5 / 16.5	−6.7 " "	9.8 / 9.8 / 9.8	16.5 / 17.0 / 17.4	−8.2 " −8.3	8.3 / 8.8 / 9.1	17.3	−9.6	7.7 / 7.9	3.5 / 3.9	+3.2 "	6.7 / 7.1
2	16.0	−7.3	8.7	20.0	−7.5	12.5	16.0	"	9.3	16.8	"	8.5	17.5 / 17.6 / 17.6	" " "	8.0 / 8.0 / 8.4	4.3 / 4.4	+3.1 +3.0	7.4 / 7.4
3	14.9	−7.4	7.5	18.0	"	10.5	15.1	−6.8	8.3	16.0	"	7.7	17.6 / 17.0	−9.7 "	7.9 / 7.3	4.5 / 4.5	+2.9 "	7.4 / 7.4
4	12.0	−7.5	4.5	16.5	"	9.0	13.8	"	7.0	14.5	−8.4	6.1	16.5	"	6.8	4.0	+2.8	6.8
5	10.4	"	2.9	14.3	−7.4	6.9	11.4	−6.9	4.5	12.7	"	4.3	13.6	"	3.9	3.2	+2.6	5.8
6	9.8 / 9.0	−7.6 "	2.2 / 1.4	13.5	"	6.1	10.2 / 9.5	" −7.0	3.3 / 2.5	12.0	−8.5	3.5	13.4 / 13.2	−9.8 "	3.6 / 3.4	3.5	+2.5	6.0
7	9.0 / 10.0	−7.7 "	1.3 / 2.3	7.2 / 7.5	−7.3 "	−0.1 / 0.2	9.0 / 10.0	−7.1 −7.2	1.9 / 2.8	11.6 / 11.2	" "	3.1 / 2.7	13.0 / 13.0	" "	3.2 / 3.2	3.3 / 3.0	+2.4 "	5.7 / 5.4
8	11.0	"	3.3	7.8 / 9.6	" "	0.5 / 2.3	10.2	−7.3	2.9	11.2 / 11.5	−8.6 "	2.6 / 2.9	13.0 / 13.0	" "	3.2 / 3.2	3.0 / 2.8	+2.3 "	5.3 / 5.1
9	13.5	"	5.8	11.9	−7.2	4.7	12.0	"	4.7	12.4	−8.7	3.7	13.1	"	3.3	2.9 / 3.0	" +2.2	5.2 / 5.2
10	16.1	−7.8	8.3	14.5	"	7.3	14.9	−7.4	7.5	14.5	−8.8	5.7	16.7	"	6.9	3.5	"	5.7
11	19.3 / 20.0	" "	11.5 / 12.2	17.0	−7.1	9.9	17.5	"	9.9	17.5	−8.9	8.6	19.1	"	9.3	4.3	"	6.5
Noon	20.8 / 20.8	" "	13.0 / 13.0	15.5 / 19.0	−7.0 "	11.5 / 12.0	19.2 / 19.5	−7.5 "	11.7 / 12.0	19.5 / 20.0	−9.0 "	10.5 / 11.0	19.5	−9.9	9.6	5.6	+2.1	7.7
1	20.6 / 20.5	" "	12.8 / 12.7	19.0 / 18.9	−6.9 "	12.1 / 12.0	20.0 / 20.5	" "	12.5 / 13.0	20.5 / 20.5	−9.1 "	11.4 / 11.4	20.0	"	10.1	7.2	+2.0	9.2
2	19.0	−7.9	11.1	18.5	−6.8	11.7	20.0 / 19.6	−7.6 "	12.4 / 12.0	20.0 / 20.5	" −9.2	11.5 / 11.3	1.0 / 3.5	+9.6 +7.2	10.6 / 10.7	8.0	+1.8	9.8
3	17.0	"	9.1	16.3	"	9.5	19.0	"	11.4	20.2	"	11.0	5.2 / 5.2	+4.9 "	10.1 / 10.1	8.4 / 8.5	+1.6 "	10.0 / 10.1
4	16.0	"	8.1	14.4	"	7.6	17.4	−7.7	9.7	18.9	−9.3	9.6	5.0 / 4.3	+4.8 "	9.8 / 9.1	8.5 / 8.5	+1.4 "	9.9 / 9.9
5	12.0 / 11.4	" "	4.1 / 3.5	11.0	"	4.2	15.4	"	7.7	17.0	"	7.7	3.9	+4.6	8.5 / 8.0	8.5 / "	+1.3 "	9.8 / 9.3
6	11.0 / 11.0	" "	3.1 / 3.1	10.0	"	3.2	13.0	−7.8	5.2	15.2	−9.4	5.8	3.2	+4.4	7.6	4.8	+1.2	6.0
7	11.0 / 11.4	" "	3.1 / 3.5	8.7	"	2.2 / 1.9	12.0 / 11.0	−7.9 "	4.1 / 3.1	13.5	"	4.1	1.6 / 1.0	+4.2 "	5.8	5.6	+1.1	6.7
8	12.0	"	4.1	8.0 / 10.3	" "	1.2 / 10.3	10.3 / 11.0	−8.0 "	2.3 / 3.0	13.2 / 13.1	−9.5 "	3.7 / 3.6	0.3 / 0.4	+4.0 "	4.3 / 4.4	4.6	+1.0	5.6
9	14.5	"	6.6	11.1	−6.7	4.4	12.5	−8.1	4.4	13.1 / 13.5	" "	3.6 / 4.0	0.6 / 0.5	+3.8 "	4.4 / 4.3	3.1 / 3.1	+0.9 "	4.0 / 4.0
10	17.5	"	9.6	14.0	"	7.3	14.5	"	6.4	13.7	−9.6	4.1	0.8 / 0.5	+3.6 "	4.4 / 4.1	3.1 / 3.1	+0.8 +0.7	3.9 / 3.8
11	19.7 / 20.3	−7.8 "	11.9 / 12.5	16.2	"	9.5	15.9	"	7.8	14.7	"	5.1	0.2 / 1.2	+3.4 "	3.6 / 4.6	3.1 / 3.5	+0.6 +0.5	3.7 / 4.0
Midn't	1.4	+12.2	13.6	16.5	"	9.8	16.1	−8.2	7.9	16.2	"	6.6	1.7	+3.3	5.0	4.1	+0.4	4.5

Nov. 17. The scale reads up to 20, hence the reading 1.4 at midnight is equivalent to 21.4.
Nov. 21. At 1 P. M. the upper limit of the scale was reached, the index was changed afterwards.

RECORD AND REDUCTION OF THE TIDES. 11

Series I.—Tidal Observations from October 10, 1853, to December 28, 1853.

Hourly observations on the pulley-gauge. Adopted reading of mean level 7.0, expressed in units of the scale. Increasing numbers indicate rise of water.

November, 1853.

Mean solar hour.	23d.	Red. to level.	Ref. obs.	24th.	Red. to level.	Ref. obs.	25th.	Red. to level.	Ref. obs.	26th.	Red. to level.	Ref. obs.	27th.	Red. to level.	Ref. obs.	28th.	Red. to level.	Ref. obs.
1	4.7	+0.4	5.1	6.7 / 7.4 / 7.8	−3.0 / −3.2 / −3.3		3.7 / 4.2 / 4.5	10.0 / 10.2 / 10.4	−6.3 " −6.4	3.7 3.9 4.0	14.8 14.5 14.6	−10.3 −10. "	4.5 4.2 4.3	14.7	−10.4	4.3	17.5	−12.0	5.5
2	5.6	+0.2	5.8	8.1	−3.4		4.7	10.7	−6.5	4.2	14.7	−10.4	4.3	13.7 / 13.3	" "	3.3 2.9	15.3 14.9	−12.1 "	3.2 2.8
3	6.2	0.0	6.2	9.1	−3.5		5.6	11.2	−6.6	4.6	15.2	−10.5	4.7	13.1 / 13.0	−10.3 "	2.8 2.7	14.7 14.5	−12.2 "	2.5 2.3
4	7.0	−0.2	6.8	10.0	−3.6		6.4	12.5	−6.7	5.8	15.5	−10.6	4.9	13.2	−10.4	2.8	14.5 / 14.8	−12.3 "	2.2 2.5
5	7.0	−0.3	6.7	10.5	−3.7		6.8	13.6	−6.9	6.7	16.5	"	5.9	16.0	"	5.6	15.6	−12.4	3.2
6	7.5 / 7.5	−0.5 / −0.6	7.0 6.9	11.1	−3.8		7.3	15.0 / 15.6	−7.0 "	8.0 8.6	18.0	−10.7	7.3	17.3	"	6.9	18.2	−12.5	5.7
7	7.6 / 7.8	−0.7 / −0.8	6.9 7.0	11.6 11.9	−3.9 "		7.7 8.0	15.7 16.0	−7.1 −7.2	8.6 8.8	19.4	"	8.7	19.8	−10.5	9.3	20.3	−12.6	7.7
8	7.9 / 7.8	−0.9 / −0.9	7.0 6.9	12.5 11.6	−4.0 −4.1		8.5 7.5	15.6	−7.3	8.3	20.0 20.2	−10.8 "	9.2 9.4	20.9 21.2	−10.6 "	10.3 10.6	22.6 23.8	−12.7 "	9.9 11.1
9	7.7 / 7.7	−1.0 "	6.7 6.7	11.0	−4.2		6.8	15.2	−7.4	7.8	20.0 19.5	" "	9.2 8.7	21.5 21.6	−10.7 "	10.8 10.9	24.5 24.7	−12.8 "	11.7 11.9
10	7.7 / 7.7	−1.1 "	6.6 6.6	9.5	−4.4		5.1	15.7	−7.6	8.1	19.0	−10.7	8.3	21.3 20.9	−10.8 "	10.5 10.1	24.8 24.3	" "	12.0 11.5
11	8.0	−1.2	6.8	9.0 / 8.8	−4.5 "	4.5 4.3	16.0	−7.8	8.2	17.8	"		7.1	20.9	−10.9	10.0	23.8	−12.9	10.9
Noon	8.6	−1.3	7.3	8.9 / 8.9	−4.6 "	3.4 4.3	15.5 / 15.5	−8.0 −8.1	7.5 7.4	16.3	"		5.6	19.2	"	8.3	22.7	"	9.8
1	9.6	−1.5	8.1	9.9	−4.7	5.2	15.4 / 15.2	−8.2 −8.3	7.2 6.9	15.7 15.4	" "		5.0 4.7	18.2	−11.0	7.2	21.4	−13.0	8.4
2	10.4	−1.6	8.8	10.3	−4.8	5.5	15.8	−8.4	7.4	15.1 15.0	" "		4.4 4.3	17.1	"	6.1	18.4	"	5.4
3	11.4	−1.8	9.6	11.4	−4.9	6.5	16.0	−8.6	7.4	15.1 15.4	" "		4.4 4.7	15.2 14.9	−11.1 "	4.1 3.8	17.1 15.8	" "	4.1 2.8
4	11.8 / 12.0	−2.0 / −2.1	9.8 9.9	12.6	−5.0	7.6	16.4	−8.8	7.6	15.5	−10.6	4.9	14.6 / 16.9	−11.2 "	3.4 5.7	15.7 15.6	" "	2.7 2.6
5	12.0 / 12.0	−2.2 / −2.3	9.8 9.7	13.5 13.9	−5.1 −5.2	8.4 8.7	18.1 19.1	−9.0 −9.1	9.1 10.0	17.0	"		6.4	17.5	−11.3	6.2	15.9 16.4	" "	2.9 3.4
6	12.2 / 11.8	" −2.4	9.9 9.4	14.6 14.9	−5.3 −5.4	9.3 9.5	19.6 19.7	−9.3 −9.4	10.3 10.3	19.1	"		8.5	19.3	−11.4	7.0	17.7	"	4.7
7	11.4	"	9.0	14.4 / 14.4	−5.5 −5.6	8.9 8.8	19.8 20.0	−9.5 −9.7	10.3 10.3	20.0 20.5	" "		9.4 9.9	20.9 22.0	−11.5 "	9.4 10.5	20.0	"	7.0
8	10.0	−2.5	7.5	13.7	−5.7	8.0	19.4	−9.8	9.6	20.4 20.4	−10.5 "		9.9 9.9	22.1 22.8	−11.6 "	10.5 11.2	22.0	"	9.0
9	9.0	−2.6	6.4	12.8	−5.9	6.9	18.5	−9.9	8.6	20.4 20.1	" "		9.9 9.6	22.9 22.9	−11.7 "	11.2 11.2	23.7 23.9	" "	10.7 10.9
10	8.2	−2.7	5.5	12.5	−6.0	6.5	18.0	−10.0	8.0	19.8	"		9.3	22.6 22.1	−11.8 "	10.8 10.3	23.9 24.0	" "	10.9 11.0
11	7.7 / 7.4	−2.8 / −2.9	4.9 4.5	11.5	−6.1	5.4	16.0	−10.1	5.9	17.7	"		7.2	21.6	−11.9	9.7	23.5 21.1	" "	12.5 8.1
Midn't	7.2	−3.0	4.2	10.9	−6.2	4.7	15.0	−10.3	4.7	16.2	−10.4	5.8	19.7	−12.0	7.7	20.8	"	7.8

Nov. 26. From 11 A. M. of this day two readings are given for each half hour; the mean of the two observations has been inserted above. The two corresponding readings agree generally within a few tenths, the difference being due to the effect of the small waves.

Nov. 28. Orders were given to observe and record careful soundings by lead line at the tide hole every day at twelve o'clock.

RECORD AND REDUCTION OF THE TIDES.

SERIES I.—TIDAL OBSERVATIONS FROM OCTOBER 10, 1853, TO DECEMBER 28, 1853.

Hourly observations on the pulley-gauge. Adopted reading of mean level 7.0, expressed in units of the scale. Increasing numbers indicate rise of water.

November, 1853. December, 1853.

Mean solar hour.	29th.	Red. to level.	Ref. obs.	30th.	Red. to level.	Ref. obs.	1st.	Red. to level.	Ref. obs.	2d.	Red. to level.	Ref. obs.	3d.	Red. to level.	Ref. obs.	4th.	Red. to level.	Ref. obs.
1	18.7	−13.0	5.7	13.1	−5.8	7.3	15.0	−6.6	8.4	18.0 17.6	−8.7 "	9.3 8.9	19.3 18.8	−9.9 "	9.4 8.9	18.0 18.6 19.2	−10.8 " −10.9	7.2 7.8 8.3
2	15.8	−12.9	2.9	9.3	"	3.5	12.4	"	5.8	15.5	−8.8	6.7	17.8	"	7.9	19.2 19.1	" "	6.3 8.2
3	14.7 14.1	" "	1.8 1.2	7.2 7.0	" "	1.4 1.2	9.2	−6.7	2.5	12.1	−8.9	3.2	16.0	"	6.1	18.7	−11.0	7.7
4	13.8 13.5	−12.8 "	1.0 0.7	6.2 5.9	" "	0.4 0.1	8.2	−6.8	1.4	10.1	−9.0	1.1	14.1	"	4.2	17.1	"	6.1
5	13.6 14.2	" −12.7	0.8 1.5	6.8 8.2	" "	1.0 2.4	6.0 5.6	" "	−0.8 −1.2	8.0 7.3	−9.1 "	−1.1 −1.8	11.6	"	1.7	14.0 12.5	−11.1 "	2.9 1.4
6	14.7	"	2.0	9.6	"	3.8	6.5 6.7	−6.9 "	−0.4 −0.2	7.1 9.0	−9.2 −9.3	−2.1 −0.3	10.1 9.7	" "	0.2 −0.2	11.4 11.3	−11.2 "	0.2 0.1
7	16.8	−12.6	4.2	11.7	"	5.9	7.5	−7.0	0.5	9.8	−9.4	0.4	9.5 10.0	" "	−0.4 0.1	11.5 12.0	" −11.3	0.3 0.7
8	19.7	−12.5	7.2	13.8	"	8.0	11.0	"	4.0	10.5	−9.5	1.0	11.2	"	1.3	12.0	"	0.7
9	23.5	−12.4	11.1	17.7	"	11.9	13.8	−7.1	6.7	14.6	−9.6	5.0	14.2	"	4.3	13.8	"	2.5
10	25.2 25.5	−12.3 −12.2	12.9 13.3	19.2 19.5	" "	13.4 13.7	18.1	−7.2	10.9	17.8	−9.7	8.1	16.0	"	6.1	17.3	−11.4	5.9
11	25.5 25.8	−12.1 −12.0	13.4 13.8	19.5 19.2	" "	13.7 13.4	19.3 19.8	−7.3 "	12.0 12.5	20.2 22.4	−9.8 −9.9	10.4 12.5	18.3	"	8.4	19.8	"	8.4
Noon	18.2 17.1	−5.8 "	12.4 11.8	18.3	"	12.5	20.2 20.0	−7.4 "	12.6 12.6	22.9 23.3	−10.0 "	12.9 13.3	21.6	−9.8	11.8	21.7	−11.5	10.2
1	15.2	"	9.4	16.3	"	10.5	19.8	−7.5	12.3	23.6 23.1	" "	13.6 13.1	22.5 22.6	" "	12.7 12.8	23.8 24.7	" "	12.3 13.2
2	11.7	"	5.9	13.0	−5.9	7.1	16.9	−7.6	9.3	22.0	"	12.0	22.7 21.5	−9.9 "	11.6	25.2 24.3	" "	13.7 12.8
3	8.7 8.2	" "	2.9 2.4	9.0	"	3.1	13.7	−7.7	6.0	19.8	"	9.8	20.5	10.0	10.5	23.9	"	12.4
4	7.0 7.2	−5.8 "	1.2 1.4	8.1 8.0	−6.0 "	2.1 2.0	10.1	−7.8	2.3	16.2	"	6.2	18.0	10.1	7.9	21.8	"	10.3
5	7.1 7.4	" "	1.3 1.6	7.7 7.7	" "	1.7 1.7	8.5 8.1	−7.9 "	0.6 0.2	12.9	"	2.9	15.7	"	5.6	19.8	"	8.3
6	7.8 9.2	" "	2.0 3.4	8.0 9.2	−6.1 "	1.9 3.1	7.7 7.7	−8.0 "	−0.3 −0.3	10.9 10.3	" "	0.9 0.3	14.5	10.2	4.3	16.7	"	5.2
7	10.0	"	4.2	10.2	"	4.1	8.2	−8.1	0.1	10.0 10.6	" "	0.0 0.6	11.9 11.2	10.3 "	1.6 0.9	14.5	"	3.0
8	13.3	"	7.5	11.9	−6.2	5.7	11.1	−8.2	2.9	10.6	"	0.6	10.4 10.3	10.4 "	0.0 −0.1	13.6 13.3	" "	2.1 1.8
9	14.8	"	9.0	14.4	"	8.2	12.6	−8.3	4.3	12.0	"	2.0	11.3	10.5	0.8	13.1 13.3	" "	1.6 1.8
10	15.9 16.5	" "	10.1 10.7	16.2	−6.3	9.9	15.3	−8.4	6.9	14.9	"	4.9	13.7	10.6	3.1	14.0	"	2.5
11	16.5 16.3	" "	10.7 10.5	16.6 17.0	−6.4 "	10.2 10.6	17.3 17.9	−8.5 "	8.8 9.4	17.0 18.2	" "	7.0 8.2	15.6	10.7	4.9	16.0	"	4.5
Midn't	16.4	"	10.3	17.3	−6.5	10.8	18.1	−8.6	9.5	18.9	"	8.9	18.2	10.8	7.4	17.8	−11.5	6.3

Nov. 29. Tide register corrected at noon. Sounding at noon 7 fath., 0 feet, 4 inches, register 18.2. It may be remarked that soundings are subject to uncertainty in case of any drift of the ice field in which the vessel was imbedded, and also in case the bottom be soft. Some allowance must be made for stretch of the line.

			Fath.	Feet	Inch.	Reg.	
Nov. 30.	Sounding at noon		7	2	3	18.2	(This sounding was not used, apparently not reliable.)
Dec. 1.	"	"	8	0	0	20.2	
" 2.	"	"	7	5	6	22.7	A mean correction was used for these days, as deduced from enveloping curves and the soundings.
" 3.	"	"	8	0	0	22.0	
" 4.	"	"	7	4	0	22.0	

RECORD AND REDUCTION OF THE TIDES. 13

SERIES I.—TIDAL OBSERVATIONS FROM OCTOBER 10, 1853, TO DECEMBER 28, 1853.

Hourly observations on the pulley-gauge. Adopted reading of mean level 7.0, expressed in units of the scale. Increasing numbers indicate rise of water.

December, 1853.

Mean solar hour.	5th.	Red. to level.	Ref. obs.	6th.	Red. to level.	Ref. obs.	7th.	Red. to level.	Ref. obs.	8th.	Red. to level.	Ref. obs.	9th.	Red. to level.	Ref. obs.	10th.	Red. to level.	Ref. obs.
1	19.7	−11.6	8.1	20.1	−15.3	4.8	4.7	−2.7	2.0	2.3 2.5 3.1	+1.4 +1.0 +0.8	3.7 3.5 3.9	1.8 1.2 1.4	−0.1 " −0.2	1.7 1.1 1.2	7.5 7.3	−1.8 "	5.7 5.5
2	20.2 20.5 20.8	" −11.7 "	8.6 8.8 9.1	21.6 22.1	−15.8 −16.1	5.8 6.0	5.2	−3.1	2.1	3.7	+0.7	4.4	1.6	"	1.4	7.1 7.2	−1.9 "	5.2 5.3
3	20.4 19.9	−11.8 "	8.6 8.1	22.5 22.4	−16.4 −16.6	6.1 5.8	7.9	−3.5	4.4	4.7	+0.5	5.2	2.7	−0.3	2.4	7.5	−2.0	5.5
4	19.2	−11.9	7.3	22.5 22.5	−16.9 −17.1	5.6 5.4	9.0	−3.9	5.1	5.6	−0.3	5.9	3.6	"	3.3	8.2	"	6.2
5	17.9	−12.0	5.9	22.1	−17.3	4.8	10.0	−4.4	5.6	7.0 8.0	0.0 −0.7	7.0 7.3	7.0	−0.4	6.6	9.5	−2.1	7.4
6	17.1	−12.1	5.0	21.2	−17.6	3.6	11.0 12.1	−5.0 −5.2	6.0 6.9	9.1 9.8	−1.5 −2.2	7.6 7.6	9.1	"	8.7	11.2	−2.2	9.0
7	14.7 14.3	−12.2 "	2.5 2.1	21.3 20.5	−17.9 −18.1	3.4 2.4	11.9 12.5	−5.4 −5.7	6.5 6.8	9.9	−3.0	6.9	10.3	−0.5	9.8	12.3	"	10.1
8	13.6 13.6	−12.3 "	1.3 1.3	20.6 20.7	−18.3 −18.4	2.3 2.3	12.4 12.5	−5.9 −6.1	6.5 6.4	11.3	−4.5	6.8	11.3 11.8	" −0.6	10.8 11.2	13.0 13.1	−2.3 "	10.7 10.8
9	13.7 14.4	−12.4 "	1.3 2.0	21.1	−18.6	2.5	12.3 11.7	−6.3 −6.5	6.0 5.2	11.4	−6.0	5.4	11.6 11.2	" −0.7	11.0 10.5	13.3 13.5	−2.4 "	10.9 11.1
10	15.2	−12.5	2.7	21.6	−18.9	2.7	12.2	−6.8	5.4	12.4	−7.5	4.9	10.7	"	10.0	13.3	−2.5	10.8
11	17.8	−12.6	5.2	22.7	−19.3	3.4	11.2	−7.3	3.9	13.2	−9.0	4.2	9.7	−0.8	8.9	11.8	"	9.3
Noon	19.0	−12.7	6.3	24.7	−19.6	4.1	11.1	−7.7	3.4	13.6 14.1	−10.5 −11.2	3.1 2.9	8.0	−0.9	7.1	9.8	−2.6	7.2
1	20.9	−12.9	8.0	26.7	−19.9	6.8	17.1	−13.1	4.0	15.0	−12.0	3.0	4.5 4.7	" −1.0	3.6 3.7	9.2 8.6	−2.7 −2.8	6.5 5.8
2	23.4 23.4	−13.1 −13.2	10.4 10.2	28.3 29.5	−20.3 −20.4	8.0 9.1	19.7	−13.6	6.1				4.7	"	3.7	8.2 8.3	−2.9 "	5.3 5.4
3	24.1 24.3	−13.3 −13.2	10.8 11.1	30.3 31.0	−20.6 −20.8	9.7 10.2	21.3	−14.2	7.1				5.2	−1.1	4.1	8.7 9.2	−3.0 −3.1	5.7 6.1
4	24.1	−13.5	10.6	31.3 31.5	−21.0 −21.2	10.3 10.3	24.2	−14.7	9.5				7.2	−1.2	6.0	9.7	"	6.6
5	23.2	−13.7	9.5	31.5 30.5	−21.3 −21.5	10.2 9.0	4.9	+4.8	9.7				10.2	"	9.0	10.2	−3.2	7.0
6	21.6	−13.9	7.7	28.5	−21.7	6.8	5.3 5.7	+4.4 +4.2	9.7 9.9				11.2	−1.3	9.9	11.2	−3.3	7.9
7	18.9	−14.1	4.8	27.5	−22.0	5.5	6.4 6.8	+4.0 +3.8	10.4 10.6	Readings omitted.			12.8 13.0	" −1.4	11.5 11.6	11.9 12.2	−3.5 "	8.4 8.7
8	17.2	−14.2	3.0	24.7 24.5	−22.4 −22.5	2.3 2.0	7.0 6.7	+3.6 +3.4	10.6 10.1		Corrections uncertain.	Rope slipping on the wheel.	13.2 13.1	" −1.5	11.8 11.6	12.9 12.7	−3.6 −3.7	9.3 9.0
9	16.1 15.5	−14.4 −14.5	1.7 1.0	24.7 24.9	−22.7 −22.9	2.0 2.0	6.0	+3.2	9.2				12.4	"	10.9	13.0 13.0	−3.8 −3.9	9.2 9.1
10	15.8 16.6	−14.6 −14.7	1.2 1.9	25.0 24.7	−23.1 −23.3	1.9 1.4	4.2	+2.7	6.9				11.1	−1.6	9.5	12.6	−4.0	8.6
11	17.1	−14.8	2.3	24.7 24.7	−23.5 −23.7	1.2 1.0	2.9	+2.2	5.1				9.8	"	8.2	11.2	−4.1	7.1
Midn't	18.9	−15.0	3.9	25.1	−23.9	1.2	2.0	+1.8	3.8				8.2	−1.7	6.5	9.5	−4.2	5.3

```
                 Fath. Feet. Inch.  Reg.
Dec. 5. Sounding at noon  7   1    0    20, changed to 19. ⎫ On these days the corrections deduced
 "   6.     "       "     6   5    0    25                 ⎬ by enveloping curves or soundings
 "   7.     "       "     6   4    0    11.1, changed to 16. ⎭ agree very well.
 "   8.     "       "     6   3    7    13.6
 "   9.     "       "     6   4    0     7.6 (The mean of the two readings is 8.0,
 "  10.     "       "     6   4    6     9.7 ( "    "     "    "     "     "   9.8.)
```

From the 9th to the 18th of December the corrections deduced from curves and soundings differ by a constant of nearly 4 feet. The differences are partly due to imperfect soundings, partly to sudden changes of the pulley-gauge (see readings between noon and 1 P. M. on the 9th). The heights given, as corrected by the soundings and curves, are, during this period, of little value, the times being less affected. The soundings were increased by 4 feet, equal to a reading of the mean level of 32.6.

RECORD AND REDUCTION OF THE TIDES.

SERIES I.—TIDAL OBSERVATIONS FROM OCTOBER 10, 1853, TO DECEMBER 28, 1853.

Hourly observations on the pulley-gauge. Adopted reading of mean level 7.0, expressed in units of the scale. Increasing numbers indicate rise of water.

December, 1853.

Mean solar hour.	11th.	Red. to level.	Ref. obs.	12th.	Red. to level.	Ref. obs.	13th.	Red. to level.	Ref. obs.	14th.	Red. to level.	Ref. obs.	15th.	Red. to level.	Ref. obs.	16th.	Red. to level.	Ref. obs.
1	8.2	—4.3	3.9	12.2	—8.8	3.4	21.4 / 20.7	—14.0 / —14.1	7.4 / 6.6	13.9	—7.6	6.3	15.4	—8.5	6.9	23.0 / 22.7	—10.6 / —10.8	12.4 / 11.9
2	7.4	—4.4	3.0	11.3	—9.1	2.0	20.3	—14.3	6.0	12.9	"	5.3	13.7	"	5.2	20.4	—11.1	9.3
3	6.8 / 6.3	—4.5 / —4.6	2.3 / 1.7	10.2 / 10.0	—9.3 / —9.5	0.9 / 0.5	19.8 / 18.9	—14.5 / —14.6	5.3 / 4.3	11.2 / 10.2	—7.7 "	3.5 / 2.5	12.2	—8.6	3.6	18.5	—11.4	7.1
4	6.5 / 7.5	—4.7 / —4.8	1.8 / 2.7	10.0 / 10.2	—9.7 / —9.8	0.3 / 0.4	18.2 / 18.7	—14.7 / —14.8	3.5 / 3.9	10.0 / 10.0	" "	2.3 / 2.3	11.5 / 11.3	" "	2.9 / 2.7	16.0 / 15.3	—11.7 / —11.8	4.3 / 3.5
5	8.7	—4.9	3.8	10.4	—10.0	0.4	19.4	—14.9	4.5	10.0 / 10.2	" "	2.3 / 2.5	11.1 / 11.4	—8.7 "	2.4 / 2.7	15.4 / 15.9	—12.0 / —12.2	3.4 / 3.7
6	11.8	—5.0	6.8	12.9	—10.2	2.7	20.8	—15.1	5.7	10.6	—7.8	2.8	11.7 / 12.2	" —8.8	3.0 / 3.4	16.1	—12.3	3.8
7	12.4	—5.1	7.3	15.5	—10.4	5.1	22.5	—15.3	7.2	12.1	"	4.3	12.9	"	4.1	17.2	—12.6	4.6
8	13.7 / 14.2	—5.2 / —5.3	8.5 / 8.9	18.0	—10.7	7.3	24.7	—15.5	9.2	15.1	—7.9	7.2	15.2	"	6.4	19.1	—12.9	6.2
9	14.7 / 15.0	—5.4 / —5.5	9.3 / 9.5	19.7	—10.9	8.8	25.7	—15.7	10.0	17.2	"	9.3	17.8	—8.9	8.9	21.2	—13.4	7.8
10	15.0 / 14.7	" —5.6	9.5 / 9.1	20.4 / 20.8	—11.1 / —11.2	9.3 / 9.6	26.7 / 27.0	—15.9 / —16.0	10.8 / 11.0	19.4 / 20.2	—8.0 "	11.4 / 12.2	19.6	"	10.7	24.0	—13.7	10.3
11	14.3	"	8.7	21.2 / 20.2	—11.3 / —11.4	9.9 / 8.8	27.0 / 27.1	—16.1 / —16.2	10.9 / 10.9	20.5 / 20.2	" "	12.5 / 12.2	20.5	"	11.6	26.0	—14.1	11.9
Noon	13.4	—5.7	7.7	20.2	—11.5	8.7	26.8 / 12.0	—16.3 / ?	10.5 / ---	19.6	—8.1	11.5	21.0 / 22.0	—9.0 "	12.0 / 13.0	27.3 / 27.5	—14.5 / —14.7	12.8 / 12.6
1	11.2	—5.9	5.3	20.2	—11.7	8.5	11.9	---	---	18.2	"	10.1	22.3 / 21.2	—9.1 "	13.2 / 12.1	27.3 / 26.8	—14.8 / —15.0	12.5 / 11.8
2	9.4 / 8.7	—6.1 / —6.2	3.3 / 2.5	18.5	—11.9	6.6	10.9	?	---	16.2	"	8.1	20.1	—9.2	10.9	26.0	—15.1	10.9
3	8.3 / 8.3	—6.4 / —6.5	1.9 / 1.8	16.8	—12.1	4.7	10.8	—7.1	3.7	13.8	—8.2	5.6	17.6	—9.3	8.3	23.1	—15.3	7.8
4	8.7	—6.7	2.0	15.9 / 15.2	—12.3 / —12.4	3.6 / 2.8	10.2 / 10.0	—7.2 "	3.0 / 2.8	11.5 / 11.1	" "	3.3 / 2.9	16.2	"	9.4	21.0	—15.5	5.5
5	9.7	—6.9	2.8	15.6	—12.5	3.1	10.3 / 10.3	" "	3.1 / 3.1	10.8 / 10.8	" "	2.6 / 2.6	15.6 / 15.4	" "	6.1 / 5.9	18.3	—15.6	2.7
6	11.2	—7.1	4.1	17.7	—12.7	5.0	10.7	—7.3	3.4	10.7 / 10.3	—8.3 "	2.4 / 2.6	15.1 / 15.2	—9.6 "	5.5 / 5.6	17.6 / 17.5	—15.8 "	1.8 / 1.7
7	12.3	—7.3	5.0	19.7	—12.9	6.8	12.2	"	4.9	11.2	"	2.9	15.7	—9.7	6.0	17.5 / 17.5	—15.9 "	1.6 / 1.6
8	13.7	—7.6	6.1	21.0	—13.1	7.9	13.0	"	5.7	13.7	"	5.4	16.8	—9.8	7.0	17.7	—16.0	1.7
9	15.2 / 15.7	—7.8 / —7.9	7.4 / 7.8	21.9	—13.3	8.6	15.1 / 15.7	—7.4 "	7.7 / 8.3	15.6	"	7.3	18.8	—9.9	8.9	19.7	—16.1	3.6
10	15.8 / 15.8	—8.1 / —8.3	7.7 / 7.5	22.8 / 23.2	—13.5 / —13.6	9.3 / 9.6	16.0 / 16.0	" "	8.6 / 8.6	16.9 / 17.2	—8.4 "	8.5 / 8.8	21.2	—10.1	11.1	21.5	—16.2	5.3
11	15.3	—8.4	6.9	23.3 / 23.4	—13.7 / —13.8	9.6 / 9.6	16.0 / 15.7	—7.5 "	8.5 / 8.2	17.3 / 17.2	" "	8.9 / 8.8	22.4 / 22.9	—10.3 / —10.4	12.1 / 12.5	23.1	—16.3	6.8
Midn't	13.7	—8.6	5.1	23.4	—13.9	9.5	15.2	"	7.7	17.1	"	8.7	23.0	—10.5	12.5	23.8	—16.4	7.4

```
                                    Fath.  Feet.  Inch.   Reg.
Dec. 11. Sounding at noon    6      2      9     13.3
 "   12.      "        "     6      4      0     19.9
 "   13.      "        "     7      1      3     13.3 (Changed from 26.8)
 "   14.      "        "     7      2      2     19.4
 "   15.      "        "     7      4      0     21.2
 "   16.      "        "     7      3      6     27.6
```

RECORD AND REDUCTION OF THE TIDES.

SERIES I.—TIDAL OBSERVATIONS FROM OCTOBER 10, 1853, TO DECEMBER 28, 1853.

Hourly observations on the pulley-gauge. Adopted reading of mean level 7.0, expressed in units of the scale. Increasing numbers indicate rise of water.

December, 1853.

Mean solar hour.	17th. *	Red. to level.	Ref. obs.	18th. *	Red. to level.	Ref. obs.	19th. *	Red. to level.	Ref. obs.	20th.	Red. to level	Ref. obs.	21st.	Red. to level.	Ref. obs.	22d.	Red. to level.	Ref. obs.
	24.0	—16.5	7.5	23.4	—17.6	5.8	4.1	+3.0	7.1									
1	23.7	—16.6	7.1	23.4	"	5.8	4.2	+2.9	7.1	14.5	—3.9	10.6	20.4	—10.7	9.7	20.6	—13.7	6.9
				23.4	"	5.8	5.0	+2.7	7.7									
2	23.8	—16.7	7.1	23.2	"	5.6	5.7	+2.5	8.2	15.1	—4.2	10.9	21.4	—11.0	10.4	21.6	"	7.9
				22.6	"	5.0	5.7	+2.3	8.0	15.3	—4.4	10.9	22.4	—11.1	11.3			
3	22.9	"	6.2	21.7	"	4.1	5.0	+2.2	7.2	15.2	—4.5	10.7	22.9	—11.3	11.6	22.2	"	8.5
										14.2	—4.7	9.5	22.3	—11.4	10.9	22.4	"	8.7
4	21.5	—16.8	4.7	20.2	"	2.6	3.7	+2.0	5.7	13.5	—4.8	8.7	21.3	—11.6	9.7	22.5	—13.8	8.7
																22.9	"	9.1
5	20.5	"	3.7	21.3	"	3.7	2.1	+1.8	3.9	12.2	—5.2	7.0	21.0	—11.9	9.1	22.3	"	8.5
	19.9	"	3.1															
6	19.6	—16.9	2.7	20.0	"	2.4	1.4	+1.6	3.0	10.5	—5.5	5.0	19.3	—12.2	7.1	21.7	"	7.9
	19.7	"	2.8	19.3	"	1.7	1.1	+1.4	2.5									
7	20.2	"	3.3	18.5	"	0.9	1.1	+1.3	2.4	10.1	—5.8	4.3	18.3	—12.4	6.9	20.9	"	7.1
				20.2	"	2.6	0.9	+1.1	2.0									
8	21.7	—17.0	4.7	20.4	—17.6	2.8	1.6	+1.0	2.6	9.6	—6.1	3.5	17.8	—12.6	5.2	19.3	—13.0	5.4
										7.7	—6.3	1.4	17.2	—12.7	4.5	19.4	"	5.5
9	25.0	"	8.0	4.3	?	---	2.9	+0.8	3.7	8.0	—6.4	1.6	17.4	—12.8	4.6	19.0	"	5.1
										8.7	—6.5	2.2	17.8	—12.9	4.9	19.0	"	5.1
10	27.7	—17.1	10.6	7.8	"	---	5.0	+0.6	5.6	9.5	—6.7	2.8	18.3	—13.0	5.3	19.3	"	5.4
11	28.8	"	11.7	8.3	"	---	7.1	+0.3	7.4	11.3	—7.1	4.2	19.5	—13.2	6.3	20.3	"	6.4
	29.7	—17.2	12.5															
Noon	23.0	?	---	10.9	"	---	8.2	0.0	8.2	13.1	—7.5	5.6	21.2	—13.4	7.8	21.7	—14.0	7.7
	22.6	"	---										27.8	?	---			
1	21.8	"	---	12.4	"	---	10.4	—0.2	10.2	16.9	—7.7	9.2	27.7	"	---	23.0	"	9.0
				12.5	"	---	10.7	—0.3	10.4									
2	20.0	"	---	12.5	"	---	10.7	—0.5	10.2	18.2	—7.9	10.3	24.4	—13.4	11.0	24.5	—14.1	10.4
							10.7	—0.6	10.1	18.4	—8.0	10.4						
3	18.6	"	---			---	10.6	—0.8	9.8	18.5	—8.1	10.4	24.7	—13.5	11.2	25.3	—14.2	11.1
										18.5	—8.2	10.3	25.0	"	11.5	25.0	"	10.8
4	18.5	—17.6	0.9	---		---	9.0	—1.0	8.0	17.9	—8.3	9.6	25.3	"	11.5	25.8	—14.3	11.5
	18.3	"	0.7										24.4	"	10.9	25.9	"	11.6
5	18.2	"	0.6	---		---	8.2	—1.2	7.0	17.0	—8.5	8.5	23.5	"	10.0	25.6	—14.4	11.2
	18.2	"	0.6							16.2	—8.6	7.6				25.1	"	10.7
6	18.3	"	0.7	---		---	8.5	—1.5	7.0	15.6	—8.7	6.9	22.1	"	8.6	24.3	—14.5	9.8
	18.4	"	0.8							16.0	—8.9	7.1						
7	18.3	?	---	15.2	?	---	7.7	—1.7	6.0	16.6	—9.1	7.5	20.2	"	6.7	22.5	—14.6	7.9
	18.3	"	---				7.6	—1.9	5.7									
8	18.3	"	---	15.4	"	---	7.6	—2.0	5.6	17.0	—9.4	7.6	19.3	—13.6	5.7	21.0	—14.7	6.3
	18.8	"	---				9.8	—2.2	7.6				18.3	"	4.7			
9	19.4	"	---	16.0	"	---	10.4	—2.4	8.0	17.0	—9.7	7.3	18.0	"	4.4	20.8	—14.8	6.0
													18.0	"	4.4			
10	20.8	"	---	17.6	"	---	11.6	—2.8	8.8	17.6	—10.0	7.6	18.0	"	4.4	19.8	—14.9	4.9
													18.3	"	4.7	19.4	"	4.5
11	21.6	"	---	20.2	"	---	13.6	—3.2	10.4	18.2	—10.2	8.0	19.2	"	5.6	19.2	—15.0	4.2
																19.4	"	4.4
Midn't	23.3	—17.6	5.7	23.2	—17.0	6.2	15.7?	—3.6	(12.1)	18.9	—10.4	8.5	19.5	—13.7	5.8	19.5	—15.1	4.7

```
                          Fath. Feet. Inch. Register.
Dec. 17. Sounding at noon   7    5    0    30.0 changed to 23.0.            * Results doubtful.
 "   18.     "        "     7    5    3    31.3 (=11.3).  Tide register broke down at 2½ P. M.; was re-
 "   19. No sounding taken.                               [paired and observations commenced at 7 P. M.
 "   20.     "        "
 "   21. Sounding at noon   7    3    6    21.5. Correction at noon by soundings 12.3, by curves 14.5,
 "   22.     "        "     7    3    6    21.7. Mean correction —14.0.                [mean adopted.
The heights on the 18th and 19th have been rejected.
```

Series I.—Tidal Observations from October 10, 1853, to December 28, 1853.

Hourly observations on the pulley-gauge. Adopted reading of mean level 7.0, expressed in units of the scale. Increasing numbers indicate rise of water.

December, 1853.

Mean solar hour.	23d.	Red. to level.	Ref. obs.	24th.	Red. to level.	Ref. obs.	25th.	Red. to level.	Ref. obs.	26th.	Red. to level.	Ref. obs.	27th.	Red. to level.	Ref. obs.	28th.	Red. to level.	Ref. obs.
1	20.0	—15.2	4.8	19.1	—16.0	3.1	19.5 19.3	" —16.9	2.7 2.4	1.9	+1.4	3.3 3.2	3.1 2.7	—0.4 "	2.7 2.3	7.1	—2.0	5.1
2	21.1	—15.3	5.8	19.7	"	3.7	19.3 19.6	" —17.0	2.4 2.6	1.8 1.9	"	3.2 3.3	2.3 2.7	—0.5 "	1.8 2.2	5.7	—2.1	3.6
3	22.0	—15.4	6.6	21.0	"	5.0	20.3	—17.1	3.2	2.1	+1.3	3.4	3.5	—0.6	2.9	4.4 4.1	" —2.2	2.3 1.9
4	22.7	—15.5	7.2	21.7	—15.9	5.8	21.8	—17.2	4.6	3.2	+1.2	4.4	3.0	"	2.4	4.2 4.3	" —2.3	2.0 2.0
5	23.0 23.1	—15.6 "	7.4 7.5	22.0	"	6.1	23.4	—17.3	6.1	5.6	"	6.8	4.7	—0.7	4.0	4.7	"	2.4
6	23.1 22.7	—15.7 "	7.4 7.0	23.4 23.6	" "	7.5 7.7	24.9	—17.4	7.5	6.7	+1.0	7.7	6.6	"	5.9	6.9	—2.4	4.5
7	23.0	—15.8	7.2	23.6 23.1	—15.8 "	7.8 7.3	25.3 25.5	—17.5 "	7.8 8.0	8.2 8.7	+0.9 "	9.1 9.6	8.5	—0.8	7.7	9.2	"	6.8
8	22.4	—15.9	6.5	22.9	"	7.1	25.6 25.6	—17.6 "	8.0 8.0	9.2 10.7	+0.6 "	10.6 11.5	10.6	—0.9	9.7	11.1	—2.5	8.6
9	21.9	—16.0	5.9	22.3	"	6.5	25.1	"	7.5	9.8	+0.7	10.5	11.9 12.2	—1.0 "	10.8 11.2	13.3 13.8	" "	10.8 11.3
10	20.6	—16.1	4.5	21.2	"	5.4	23.6	—17.7	5.9	8.5	"	9.2	12.1 11.6	—1.1 "	11.0 10.5	14.2 14.2	—2.6 "	11.6 11.6
11	19.4 19.3	—16.2 "	3.2 3.1	21.0 20.7	" "	5.2 4.9	22.9 22.3	—17.8 "	5.1 4.5	6.7	+0.6	7.3	11.0	—1.2	9.8	13.7	"	11.1
Noon	20.4	—16.3	4.1	20.3 21.3	—15.7 "	4.6 5.6	22.0 22.0	—17.9 "	4.1 4.1	5.6	+0.6	6.2	9.2	—1.3	7.9	11.7	—2.7	9.0
1	21.4	"	5.1	21.7	"	6.0	22.2	—18.0	4.2	4.4 4.4	" "	5.0 5.0	6.9	"	5.6	10.0	—2.9	7.1
2	22.5	"	6.2	22.5	—15.8	6.7	22.5	"	4.5	4.3 4.3	+0.5 "	4.8 4.8	5.2	—1.4	3.8	9.6	—3.3	6.3
3	23.6	—16.2	7.4	23.6	—15.9	7.6	23.2	—18.1	5.1	4.3 4.3	" "	4.8 4.8	4.5 4.4	—1.5 "	3.0 2.9	9.5	—3.8	5.7
4	24.6 25.0	" "	8.4 8.8	21.5	—16.0	8.5	23.6	—18.2	5.4	4.5	+0.4	4.9	4.4 4.4	" "	2.9 2.9	9.5 9.6	—4.5 —4.7	5.0 4.9
5	24.9 24.7	" "	8.7 8.5	25.1	—16.1	9.0	24.6	"	6.4	5.2	+0.3	5.5	4.7	—1.6	3.1	9.8 10.0	—4.9 —5.1	4.9 4.9
6	24.6	"	8.4	25.4 25.5	—16.2 "	9.2 9.3	25.3	—18.3	7.0	6.8	"	7.1	6.3	"	4.7	10.4	—5.3	5.1
7	23.7	—16.1	7.6	25.1	—16.3	8.8	25.3 25.6	" "	7.0 7.3	8.1 8.5	+0.2 "	8.3 8.7	7.8	—1.7	6.1	12.6	—5.7	6.9
8	22.3	"	6.2	24.2	—16.4	7.8	25.4	—18.4	7.0	9.1 9.1	+0.1 "	9.2 9.2	9.3	"	7.6	14.2	—6.0	8.2
9	20.1	"	4.0	23.1	—16.5	6.6	25.0	"	6.6	9.1	0.0	9.1	10.0 10.9	—1.8 "	8.2 9.1	16.3	—6.3	10.0
10	19.4	"	3.3	22.6	—16.6	6.0	24.1	—18.5	5.6	8.1	—0.1	8.0	10.8 10.4	" "	9.0 8.6	17.1 17.4	—6.9 —7.0	10.2 10.4
11	19.1 19.0	" "	3.0 2.9	22.1	—16.7	5.4	23.1	"	4.6	7.2	—0.2	7.0	10.2	—1.9	8.3	17.7 17.7	—7.1 —7.2	10.6 10.5
Midn't	19.0	—16.0	3.0	21.4	—16.8	4.6	22.4	—18.6	3.8	5.2	—0.3	4.9	9.0	—2.0	7.0	17.7	—7.3	10.4

                                    Fath. Feet. Inch. Correction.

Dec. 23. Sounding at noon 6 3 8 —16.3 mean of sounding and curves.
" 24. " " 6 4 8 —15.7 " " "
" 25. " " 6 3 0 —17.9 " " "
" 26. " " 7 1 0 + 0.6 " " "
" 27. " " 7 2 0 — 1.3 " " "
" 28. " " 7 4 6 (Ebb tide at 2½ P. M.) Correction —2.7 mean of sounding and curves; afternoon corrections from the curves. Between this date and the commencement of the second series the observations are too much affected by irregularities to be inserted.

RECORD AND REDUCTION OF THE TIDES. 17

SERIES II.—TIDAL OBSERVATIONS FROM JANUARY 28 TO APRIL 7, 1854.
Hourly observations on the pulley-gauge. Adopted reading of mean level 7.0, expressed in units of the scale. Increasing numbers indicate rise of water.

Mean solar hour.	January, 1854.										February, 1854.			
	28th.	Red. to level.	Ref. obs.	29th.	Red. to level.	Ref. obs.	30th.	Red. to level.	Ref. obs.	31st.	Red. to level.	Ref. obs.	1st.	Red. to level.	Ref. obs.	2d.	Red. to level.	Ref. obs.
	13.2	—2.6	10.6	21.9	—10.0	11.9	16.5	—7.5	9.0	14.9	—5.8	9.1				13.2	—6.2	7.0
1	12.8	—2.8	10.0	21.5	—10.1	11.4	16.2	—7.1	8.8	15.5	"	9.7	16.2	—5.8	10.4	13.5	"	7.3
										15.2	"	9.4						
2	12.0	—3.3	8.7	19.6	—10.2	9.4	15.6	—7.3	8.3	14.7	"	8.9	16.8	—5.9	10.9	14.3	"	8.1
													17.0	"	11.1			
3	11.0	—3.8	7.2	17.7	—10.3	7.4	12.8	—7.2	5.6	14.3	—5.7	8.6	17.0	"	11.1	15.6	"	9.4
													17.0	"	11.1	16.0	"	9.8
4	8.8	—4.2	4.6	16.1	—10.4	5.7	10.5	—7.1	3.4	11.7	"	6.0	16.9	—6.0	10.9	16.2	"	10.0
																16.2	"	10.0
5	6.3	—4.7	1.6	15.0	—10.5	4.5	9.2	—7.0	2.2	9.2	"	3.5	12.6	"	6.6	15.7	"	9.5
	5.9	—4.9	1.0															
6	6.2	—5.1	1.1	12.2	—10.6	1.6	7.4	—6.9	0.5	6.9	"	1.2	10.6	—6.1	4.5	14.0	"	7.8
	6.7	—5.3	1.4	11.9	"	1.3	7.4	"	0.5	5.9	—5.6	0.3						
7	8.2	—5.4	2.8	12.1	—10.7	1.4	7.5	—6.8	0.7	5.3	"	—0.3	8.5	"	2.4	10.6	"	4.4
				13.4	"	2.7	7.8	"	1.0	5.3	"	—0.3	7.9	"	1.8			
8	11.0	—5.9	5.1	14.6	—10.8	3.8	8.2	—6.7	1.5	5.3	"	—0.3	7.7	"	1.6	8.9	"	2.7
										5.3	—5.5	—0.2	7.7	"	1.6	9.1	"	2.9
9	15.6	—6.4	9.2	18.6	—10.9	7.7	10.3	—6.6	3.7	6.9	"	1.4	7.5	—6.2	1.3	8.2	"	2.0
													8.3	"	2.1	8.2	"	2.0
10	17.7	—7.2	10.5	20.7	—11.0	9.7	12.2	—6.5	5.7	10.0	"	4.5	10.0	"	3.8	8.7	"	2.5
11	19.9	—8.0	11.9	23.4	—11.1	12.3	14.3	—6.4	7.9	13.0	—5.4	7.6	13.0	"	6.8	11.2	"	5.0
Noon	21.0	—8.8	12.2	24.5	—11.2	13.3	18.3	—6.3	12.0	16.2	—5.3	10.9	16.3	—6.3	10.0	13.0	—6.1	6.9
	21.2	"	12.4	20.7	—11.1	9.7				17.3	"	12.0						
1	20.4	—8.9	11.5	19.7	—11.0	8.7	20.1	"	13.8	18.3	"	13.0	17.7	"	11.4	14.6	"	8.5
				19.2	—10.9	8.3	20.2	"	13.9	19.0	"	13.7						
2	18.8	—9.0	9.8	18.7	—10.8	7.9	19.7	—6.2	13.5	18.2	—5.4	12.8	18.4	"	12.1	16.0	"	9.9
													19.0	"	12.7	16.2	"	10.1
3	17.1	—9.1	8.0	15.6	—10.6	5.0	17.7	—6.1	11.6	16.8	"	11.4	19.2	"	12.9	16.7	"	10.6
													18.9	"	12.6	16.0	"	9.9
4	15.6	—9.2	6.4	12.0	—10.4	1.6	14.6	"	8.5	15.7	—5.5	10.2	18.2	"	11.9	16.0	"	9.9
5	13.7	—9.3	4.4	10.8	—10.2	0.6	8.2	—6.0	2.2	13.0	"	7.5	15.8	"	9.5	14.9	—6.2	8.7
6	12.2	—9.4	2.8	10.0	—10.0	0.0	6.3	"	0.3	9.7	—5.6	4.1	12.4	"	6.1	13.9	"	7.7
	12.0	"	2.6	9.7	—9.9	—0.2	6.1	"	0.1									
7	12.2	—9.5	2.7	9.2	—9.8	—0.6	5.7	"	—0.3	7.3	"	1.7	10.0	"	3.7	10.8	"	4.6
				8.6	—9.7	—1.1	5.6	—5.9	—0.3	7.2	"	1.6						
8	14.2	—9.6	4.6	7.8	—9.6	—1.8	5.5	"	—0.4	7.2	"	1.6	8.7	"	2.4	8.1	"	1.9
				7.7	—9.5	—1.8	5.7	"	—0.2	7.2	—5.7	1.5	8.5	"	2.2			
9	17.1	—9.7	7.4	8.7	—9.4	—0.7	6.7	"	0.8	7.4	"	1.7	8.4	"	2.1	6.6	"	0.4
										8.0	"	2.3	8.4	"	2.1	6.6	"	0.4
10	19.9	—9.8	10.1	10.8	—9.2	0.6	9.3	"	3.4	9.0	"	3.3	8.6	"	2.3	6.6	—6.3	0.3
																6.6	"	0.3
11	21.1	—9.9	11.2	13.4	—9.0	4.4	12.1	"	6.2	12.0	"	6.3	10.0	"	3.7	7.2	"	0.9
Midn't	21.6	—10.0	11.6	14.8	—8.7	6.1	14.0	—5.8	8.2	14.6	—5.8	8.8	12.1	—6.2	5.9	9.3	"	3.0

Jan. 27. At 11 P. M., 13.7; at 11ʰ 30ᵐ, 13.8; at 28th, 0ʰ, 13.9, high water; corrected high water 11.3.

	Fath.	Feet.	Inch.	
Jan. 28. Sounding at noon	8	0	0	Corrected reading by sounding 11.4, by curves 13.1, mean 12.2.
" 29. " "	8	1	6	Index changed to 19.6. Corrected reading by sounding 12.9, by curves 13.7, mean 13.3.
" 30. " "	8	0	6	Corrected reading by sounding 11.9, by curves 12.1, mean 12.0.
" 31. " "	...	...	...	Corrected reading by curves 10.9.
Feb. 1. " "	7	4	6	Ebb tide at 7½ A. M.
" 2. " "	7	1	0	Ebb tide at 4 P. M. (probably means ebb commences). Index 13.0.

Note to Feb. 1 and 2. The correction is derived from the soundings and curves.

3

Series II.—Tidal Observations from January 28 to April 7, 1854.

Hourly observations on the pulley-gauge. Adopted reading of mean level 7.0, expressed in units of the scale. Increasing numbers indicate rise of water.

February, 1854.

Mean solar hour.	3d.	Red. to level.	Ref. obs.	4th.	Red. to level.	Ref. obs.	5th.	Red. to level.	Ref. obs.	6th.	Red. to level.	Ref. obs.	7th.	Red. to level.	Ref. obs.	8th.	Red. to level.	Ref. obs.
1	12.0	—6.4	5.6	11.2	—6.0	5.2	10.6 10.9	—6.8 —7.0	3.8 3.9	11.7 11.7 12.2	—7.2 —7.1 "	4.5 4.6 5.1	9.5 9.3	—5.2 "	4.3 4.1	9.7 9.2	—4.4 "	5.3 4.8
2	13.0	"	6.6	13.0	—5.9	7.1	11.7	—7.1	4.6	12.6	—7.0	5.6	9.5	—5.1	4.4	9.2 9.2	—4.5 "	4.7 4.7
3	14.0 14.4	—6.5 "	7.5 7.9	14.3 14.7	" —5.8	8.4 8.9	12.9	—7.2	5.7	13.5	—6.9	6.6	10.2	—5.0	5.2	9.2 9.2	" "	4.7 4.7
4	15.0 15.0	" "	8.5 8.5	14.9 14.0	" "	9.1 9.1	13.7	—7.3	6.4	15.9 15.8	" —6.8	9.0 9.0	11.1	—4.9	6.2	9.2 9.6	" "	4.7 5.1
5	14.8	—6.6	8.2	14.9 14.8	—5.7 "	9.2 9.1	17.1	—7.4	9.7	16.1 16.0	" "	9.3 9.2	12.0	"	7.1	10.0	"	5.5
6	13.9	"	7.3	14.2	—5.6	8.6	17.5 17.5	—7.5 "	10.0 10.5	16.0 16.0	—6.7 "	9.3 9.3	13.0	—4.8	8.2	11.5	"	7.0
7	12.1	"	5.5	12.8	"	7.2	17.7 17.7	—7.6 "	10.1 10.1	16.0 16.0	—6.6 "	9.4 9.4	13.7 14.0	—4.7 "	9.0 9.3	12.6	—4.6	8.0
8	11.1 10.6	" "	4.5 4.0	11.7 9.8	—5.5 "	6.2 4.3	17.4	—7.7	9.7	16.0 16.0	—6.5 "	9.5 9.5	14.0 13.7	—4.6 "	9.4 9.1	13.4	"	8.8
9	10.1 10.2	—6.7 "	3.4 3.5	9.2 9.2	—5.4 "	3.8 3.8	16.3	—7.8	8.5	15.2	—6.4	8.8	13.3	—4.5	8.8	13.9 14.0	" "	9.3 9.4
10	10.8	"	4.1	9.8	—5.4	4.4	15.5	—8.0	7.5	14.4	—6.3	8.1	12.8	—4.4	8.4	14.1 14.1	" "	9.5 9.5
11	11.9	"	5.2	10.5	"	5.1	15.2	—8.1	7.1	13.6 13.0	" "	7.3 6.7	12.2	—4.3	7.9	13.9	"	9.3
Noon	12.6	—6.8	5.8	10.7	—5.3	5.4	15.0 13.2	—8.2 "	6.8 5.0	10.0 10.1	—6.2 "	3.8 3.9	11.8	—4.2	7.6	11.6	—4.7	6.9
1	12.9	"	6.1	11.2	—5.4	5.8	13.7 14.1	—8.1 "	5.6 6.0	10.5	—6.1	4.4	11.0 11.0	" "	6.8 6.8	10.2	"	5.5
2	14.1	—6.7	7.4	12.8	—5.5	7.3	14.4	—8.0	6.4	11.0	—6.0	5.0	10.0 10.0	" "	5.8 5.8	9.1 8.2	" "	4.4 3.5
3	14.8 15.1	" —6.6	8.1 8.5	13.5	—5.6	7.9	15.5	—7.9	7.6	11.3	"	5.3	10.0 10.0	" "	5.8 5.8	8.0 8.0	" "	3.3 3.3
4	15.2 ...	" 7.6	8.6 ...	14.2	—5.7	8.5	17.7	—7.8	9.9	12.0	—5.9	6.1	10.0 10.0	" "	5.8 5.8	8.0 ...	"	3.3 ...
5	14.6	—6.5	8.1	14.6 14.7	—5.8 "	8.8 8.9	18.3 18.4	—7.7 "	10.5 10.7	12.5	—5.8	6.7	11.3	—4.3	7.0	8.1	"	3.4
6	12.9	"	6.4	14.6	—5.9	8.7	18.0	—7.6	10.4	12.6 12.6	—5.7 "	6.9 6.9	11.4	"	7.1	9.7	"	5.0
7	11.2	—6.4	4.8	14.2	—6.0	8.2	17.2	—7.5	9.7	12.7 12.7	—5.6 "	7.1 7.1	12.2 12.5	" "	7.9 8.2	10.2	"	5.5
8	9.6	—6.3	3.3	13.5	—6.1	7.4	15.0	—7.4	7.6	12.7 12.7	—5.5 "	7.2 7.2	12.5 12.2	" "	8.2 7.9	11.1 11.2	" "	6.4 6.5
9	8.7 8.4	" —6.2	2.4 2.2	12.2	—6.2	6.0	13.7	—7.4	6.3	12.5	—5.4	7.1	12.0	—4.4	7.6	11.4 11.4	" "	6.7 6.7
10	8.4 8.6	" —6.1	2.2 2.5	11.0	—6.4	4.6	12.4	—7.3	5.1	11.3	"	5.9	11.8	"	7.4	11.4 11.3	" "	6.7 6.6
11	8.8	"	2.7	10.6 10.6	—6.5 —6.6	4.1 4.0	11.9 11.8	" "	4.6 4.5	9.5 9.0	—5.3 "	4.2 3.7	11.2	"	6.8	11.1	"	6.4
Midn't	9.7	—6.0	3.7	10.6	—6.7	3.9	11.7	—7.2	4.5	9.0	—5.2	3.8	10.7	—4.4	6.3	11.0	"	6.3

```
                              Fath   Feet   inch.
Feb. 3. Sounding at noon       6      5      6     Corrected reading by sounding 4.9, by curves 6.7, mean 5.8.
  "    4.     "       "       ...   ...   ...
  "    5.     "       "       ...   ...   ...
  "    6.     "       "        6      4      0     Corrected reading by sounding 3.4, by curves 4.3, mean 3.8.
  "    7.     "       "       ...   ...   ...
  "    8.     "       "        6      5      6     Corrected reading by sounding and curves 6.9.
```

SERIES II.—TIDAL OBSERVATIONS FROM JANUARY 28 TO APRIL 7, 1854.

Hourly observations on the pulley-gauge. Adopted reading of mean level 7.0, expressed in units of the scale. Increasing numbers indicate rise of water.

February, 1854.

Mean solar hour.	9th.	Red. to level.	Ref. obs.	10th.	Red. to level.	Ref. obs.	11th.	Red. to level.	Ref. obs.	12th.	Red. to level.	Ref. obs.	13th.	Red. to level.	Ref. obs.	14th.	Red. to level.	Ref. obs.
1	9.9	—4.6	5.3	11.4	—5.6	5.8	10.7	—6.3	4.4	11.1	—6.2	4.9	14.7 14.2	—6.7 "	8.0 7.5	19.5	—9.4	10.1
2	8.8 8.4	" "	4.2 3.8	11.0	—5.7	5.3	10.1	"	3.8	11.1	"	4.9	13.6	"	6.9	18.5	—9.6	8.9
3	8.4 8.4	" "	3.8 3.8	10.4	—5.8	4.6	8.8	"	2.5	10.2	"	4.0	12.7	—6.8	5.9	16.0	—9.8	6.2
4	8.4 8.4	" "	3.8 3.8	10.0	—5.9	4.1	8.2 7.9	" "	1.9 1.6	9.0 9.5	" "	2.8 3.3	9.7	"	2.9	14.0	—10.0	4.0
5	8.8	"	4.2	10.0	—6.0	4.0	7.5 7.5	—6.2 "	1.3 1.3	9.2 9.0	" "	3.0 2.8	8.2 8.0	" "	1.4 1.2	13.7	—10.1	3.6
6	10.0	"	5.4	10.0 9.8	—6.1 "	3.9 3.7	7.8	"	1.6	9.0 9.0	" "	2.8 2.8	7.6 6.5	—6.9 "	0.7 1.6	13.3 12.9	—10.3 —10.4	3.0 2.5
7	11.6	"	7.0	11.0	—6.2	4.8	9.1	"	2.9	9.7	"	3.5	8.9	"	2.0	12.9 13.7	—10.5 —10.6	2.4 3.1
8	12.8	"	8.2	12.8	—6.3	6.5	10.3	—6.1	4.2	11.2	"	5.0	10.9	"	4.0	14.7	—10.7	4.0
9	13.5	"	8.9	15.0 15.2	—6.4 "	8.6 8.8	12.6	"	6.5	13.2	"	7.0	13.0	—7.0	6.0	16.2	—10.8	5.4
10	14.1 14.3	" "	9.5 9.7	15.7 15.7	—6.5 "	9.2 9.2	15.5	"	9.4	15.3 16.6	" "	9.1 10.4	16.0 16.9	" "	9.0 9.9	18.7	—10.9	7.8
11	14.2 13.8	" "	9.6 9.2	15.7 15.0	" "	9.2 8.5	16.1 16.4	" "	10.0 10.3	17.7 17.4	" "	11.5 11.2	19.1 19.6	" "	12.1 12.6	20.8	—11.1	9.7
Noon	13.6	—4.5	9.1	14.0	—6.6	7.4	16.5 16.1	—6.0 "	10.5 10.1	17.1	—6.3	10.8	19.5 19.5	—7.1 —7.2	12.4 12.3	21.7 ---	—11.2 "	10.5 ---
1	13.5	"	9.0	13.9	"	7.3	15.7	"	9.7	16.0	"	9.7	19.5	—7.3	12.2	24.8 24.3	" "	13.6 13.1
2	12.7	"	8.2	13.4	"	6.8	13.6	"	7.6	14.7	"	8.4	17.9	—7.5	10.4	23.2	—11.3	11.9
3	11.2	—4.6	6.6	10.7	"	4.1	11.4	"	5.4	12.3	—6.4	5.9	16.7	—7.7	9.0	21.5	—11.4	10.1
4	9.0 8.3	—4.7 "	4.3 3.6	8.7 8.0	—6.5 "	2.2 1.5	9.5 8.0	" "	3.5 2.0	10.0	"	3.6	16.3 14.5	—7.9 —8.0	8.4 6.5	19.5	"	8.1
5	8.0 8.7	" —4.8	3.3 3.9	8.0 8.0	" "	1.5 1.5	7.8 8.0	" "	1.8 2.0	9.7	"	3.3	12.9 12.3	—8.1 —8.2	4.8 4.1	16.7	—11.5	5.2
6	9.3	"	4.5	8.0 8.0	" "	1.5 1.5	8.0	"	2.0	8.9 8.8	—6.5 "	2.4 2.3	13.1	—8.3	4.8	14.7 14.3	" "	3.2 2.8
7	9.7	—4.9	4.8	8.5	"	2.0	9.6	"	3.6	8.8 8.9	" "	2.3 2.4	15.2	—8.4	6.8	14.2 14.2	—11.6 "	2.6 2.6
8	11.0	—5.0	6.0	9.8	—6.4	3.4	10.3	—6.1	4.2	9.7	"	3.2	16.3	—8.5	7.8	14.6	"	3.0
9	11.7	—5.1	6.6	11.2	"	4.8	11.4	"	5.3	9.9	—6.6	3.3	16.3	—8.7	7.6	15.1	"	3.5
10	12.4	—5.2	7.2	12.3	"	5.9	12.1	"	6.0	12.6 14.3	" "	6.0 7.7	18.7 19.1	—6.8 —8.9	9.0 10.2	15.8	—11.7	4.1
11	12.3 12.6	—5.3 —5.4	7.0 7.2	12.7 12.7	" "	6.3 6.3	12.9 13.3	" "	6.8 7.2	15.0	"	8.4	19.6 19.5	—9.0 —9.1	10.6 10.4	16.6	"	4.9
Midn't	12.4	—5.5	6.9	12.9	—6.3	6.6	13.8	"	7.7	15.1	—6.7	8.4	19.5	—9.2	10.3	17.4	—11.8	5.6

Feb. 9. No sounding.
    " 10. Sounding at noon   Fath. 7   Feet 2   Inch. 0   Corrected by sounding —6.6, by curves —6.6.
    " 11.   "   "   --- --- ---
    " 12.   "   "   7   4   6   Ebb tide at 6 P.M. Corrected by sounding —6.6, by curves —6.0, mean —6.3.
    " 13.   "   "   8   1   6   Corrected by sounding —6.6, by curves —7.6, mean —7.1.
    " 14. No sounding.

Series II.—Tidal Observations from January 28 to April 7, 1854.

Hourly observations on the pulley-gauge. Adopted reading of mean level 7.9, expressed in units of the scale. Increasing numbers indicate rise of water.

February, 1854.

Mean solar hour.	15th.	Red. to level.	Ref. obs.	16th.	Red. to level.	Ref. obs.	17th.	Red. to level.	Ref. obs.	18th.	Red. to level.	Ref. obs.	19th.	Red. to level.	Ref. obs.	20th.	Red. to level.	Ref. obs.
1	23.6 23.6 23.5	—11.8 " "	11.8 11.8 11.7	22.2 22.2 22.1	—9.1 —9.2 —9.3	13.1 13.0 12.8	20.7 21.2 21.6	—9.7 —9.6 "	11.0 11.6 12.0	16.5	—8.0	8.5	17.8	—8.2	9.6	11.3	—5.4	5.9
2	23.5	—11.9	11.6	22.1	—9.4	12.7	21.8 21.8	—9.5 "	12.3 12.3	18.0 18.4	" —7.9	10.0 10.5	18.6	"	10.4	13.7	"	8.3
3	23.5	"	11.6	20.9	—9.6	11.3	21.9 21.9	—9.4 "	12.5 12.5	18.5 18.5	" "	10.6 10.6	19.4 19.6	—8.3 "	11.1 11.3	15.5 15.9	" "	10.1 10.5
4	19.7	"	7.8	19.0	—9.8	9.2	21.9	—9.3	12.6	18.2	—7.8	10.4	19.6 19.6	—8.4 "	11.2 11.2	16.2 14.9	" "	10.8 9.5
5	17.7	—12.0	5.7	16.3	—9.9	6.4	19.0	—9.2	9.8	16.3	"	8.5	19.3	"	10.9	14.0	—5.5	8.5
6	16.1	—12.1	4.0	14.7	—10.1	4.6	16.4	—9.1	7.3	15.0	—7.6	7.4	17.7	—8.5	9.2	12.7	"	7.2
7	15.3	"	3.2	13.2 12.9	—10.3 —10.4	2.9 2.5	14.3 12.9	—9.0 "	5.3 3.9	11.8	"	4.2	15.8	"	7.3	11.4	"	5.9
8	15.5	—12.2	3.3	12.7 13.7	—10.5 —10.6	2.2 3.1	12.9 13.0	—8.9 "	4.0 4.1	10.6 10.4	" "	3.0 2.8	14.2	—8.6	5.6	11.0 11.0	" "	5.5 5.5
9	17.6	—12.3	4.8	14.8	—10.7	4.1	13.5	—8.8	4.7	10.2 10.2	—7.5 "	2.7 2.7	13.5	"	4.9	11.0 11.4	" "	5.5 5.0
10	19.7	—12.4	7.3	16.0	—10.9	5.1	15.4	—8.7	6.7	10.3	"	2.8	13.3 13.3	—8.7 "	4.6 4.6	11.0 11.0	" "	5.5 5.5
11	22.3	"	9.9	17.5	—11.0	6.5	18.0 20.1	" —8.6	9.3 11.5	13.5	"	6.0	13.2 13.3	" "	4.5 4.6	11.0 11.3	" "	5.5 5.8
Noon	24.6 25.2	—12.5 "	12.1 12.7	21.2 22.1	—11.1 "	10.1 11.0	14.6	?	...	14.5	—7.4	7.1	13.6	—8.8	4.8	11.7	—5.6	6.1
1	19.5 19.6	—6.8 —6.9	12.7 12.9	24.0 23.9	—11.0 "	13.0 12.9	17.0	?	...	15.3	"	7.9	14.3	—5.5	8.8	12.2	"	6.6
2	19.8	—7.0	12.8	24.0 24.0	—10.9 "	13.1 13.1	18.3	—6.6	11.7	17.2 17.7	—7.5 "	9.7 10.2	15.6 15.9	" "	10.1 10.4	15.5	"	9.9
3	18.3	—7.2	11.1	22.5	—10.8	11.7	18.2	—6.7	11.5	17.7 17.6	—7.6 "	10.1 10.0	16.2 16.2	" "	10.7 10.7	15.5 16.0	" "	9.9 10.4
4	18.4	—7.4	11.0	21.9	—10.7	11.2	17.2	—6.9	10.3	17.5	"	9.9	15.7	"	10.2	16.3 15.7	" "	10.7 10.1
5	17.2	—7.6	9.6	18.8	—10.6	8.2	15.4	—7.1	8.3	15.9	—7.7	8.2	15.4	—5.5	9.9	15.8	"	10.2
6	14.7	—7.8	6.9	16.0	—10.5	5.5	12.9	—7.2	5.7	14.4	"	6.7	14.3	"	8.8	15.5	"	9.9
7	12.7 12.4	—8.0 —8.1	4.7 4.3	15.6	—10.4	5.2	11.2	—7.3	3.9	12.8	—7.8	5.0	12.0	"	6.5	15.5	?	---
8	12.3 13.7	—8.2 —8.3	4.1 5.4	12.9 12.5	—10.3 —10.2	2.6 2.3	8.8 8.0	—7.5 —7.6	1.3 0.4	11.6 11.1	" —7.9	3.8 3.2	10.3	"	4.8	15.5	"	---
9	15.0	—8.4	7.2	12.8	—10.1	2.7	8.0 8.0	—7.7 "	0.3 0.3	10.9 10.9	" "	3.0 3.0	8.9	—5.4	3.5	14.1	"	---
10	18.1	—8.6	9.5	14.3	—10.0	4.3	8.7	—7.8	0.9	11.0	—8.0	3.0	8.4 8.4	" "	3.0 3.0	13.7 13.6	" "	--- ---
11	19.3	—8.8	10.5	17.1	—9.9	7.2	11.0	—7.9	3.1	12.3	"	4.3	8.4 8.5	" "	3.0 3.1	13.5 13.5	" "	--- ---
Midn't	20.4	—9.0	11.4	19.7	—9.8	9.9	13.8	—8.0	5.5	14.3	—8.1	6.2	9.4	—5.3	4.1	13.5	"	---

	Fath.	Feet.	Inch.	
Feb. 15. Sounding at noon	6	1	0	Correction by sounding —12.2, by curves —12.8, mean —12.5.
" 16. " "	...	...	...	
" 17. " "	7	2	6	
" 18. " "	7	1	6	Corrected reading by sounding 6.9, by curves 7.3, mean 7.1.
" 19. " "	7	0	0	Corrected reading by sounding 5.4, by curves 4.2, mean 4.8.
" 20. No sounding.				

RECORD AND REDUCTION OF THE TIDES. 21

SERIES II.—TIDAL OBSERVATIONS FROM JANUARY 28 TO APRIL 7, 1854.

Hourly observations on the pulley-gauge. Adopted reading of mean level 7.0, expressed in units of the scale. Increasing numbers indicate rise of water.

February, 1854.

Mean solar hour.	21st.	Red. to level.	Ref. obs.	22d.	Red. to level.	Ref. obs.	23d.	Red. to level.	Ref. obs.	24th.	Red. to level.	Ref. obs.	25th.	Red. to level.	Ref. obs.	26th.	Red. to level.	Ref. obs.
1	---	---	---	10.5 10.7	—6.0 "	4.5 4.7	12.4 11.8 11.8	—7.2 —7.3 —7.5	5.2 4.5 4.3	14.4 14.2	—10.9 "	3.5 3.3	9.8	—2.1	7.7	10.6	—2.2	8.4
2	---	---	---	11.5	"	5.5	11.8 11.8	—7.4 —7.5	4.4 4.3	14.5	—11.0	3.5	7.8	"	5.7	8.8	—2.4	6.4
3	---	---	---	12.9	"	6.9	11.8 12.1	—7.7 —7.8	4.1 4.3	14.9	—11.1	3.8	5.8 5.3	—2.2 "	3.6 3.1	6.4	—2.7	3.7
4	---	---	---	14.2	"	8.2	13.0	—8.0	5.0	16.1	—11.2	4.9	5.1 15.4	—2.3 "	2.8 3.1	5.2 4.1	—3.0 —3.1	2.2 1.0
5	---	---	---	14.1	?	---	14.3	—8.2	6.1	17.2	—11.3	5.9	5.6	"	3.3	4.2 4.3	—3.2 —3.3	1.0 1.0
6	---	---	---	13.6	"	---	15.8	—8.3	7.5	18.3	—11.5	6.8	7.6	—2.4	5.2	4.7	—3.5	1.2
7	---	---	---	13.2	"	---	16.8	—8.5	8.3	19.4	—11.6	7.8	9.7	"	7.3	6.6	—3.7	2.9
8	---	---	---	12.7 15.0	" —6.2	--- 8.8	18.1	—8.6	9.5	21.7	—11.8	9.9	11.0	—2.5	8.5	9.0	—4.0	5.0
9	---	---	---	15.5 15.6	" "	9.3 9.4	18.4	—8.7	9.7	22.2 22.2	—11.9 "	10.3 10.3	12.7	"	10.2	11.0	—4.3	6.7
10	---	---	---	14.9	"	8.7	19.1 19.5	—8.8 "	10.3 10.7	22.3	—12.0	10.3	13.8 13.8	—2.6 "	11.2 11.2	13.0	—4.6	8.4
11	---	---	---	14.2	"	8.0	19.3	—8.9	10.4	22.2	—12.1	10.1	13.8	"	11.2	14.6	—4.9	9.7
Noon	---	---	---	12.8	—6.2	6.6	18.7	—9.0	9.7	20.7	—12.3	8.4	13.7	—2.7	11.0	15.4 17.0	—5.4 —5.5	10.0 11.5
1	---	---	---	12.2	"	6.0	17.9	—9.3	8.6	10.0	—1.7	8.3	11.4	—1.9	9.5	17.5 15.7	—5.6 —5.7	11.9 10.0
2	---	---	---	12.2	"	6.0	16.7	—9.6	7.1	7.6	"	5.9	9.0	—1.0	8.0	15.0	—5.8	9.2
3	---	---	---	12.2	—6.3	5.9	16.1 16.0	—9.9 —10.0	6.2 6.0	6.6 5.0	—1.8 "	4.8 3.2	6.0	—1.1	4.9	13.1	—6.0	7.1
4	---	---	---	12.5 11.0	" "	6.2 4.7	16.0 16.0	—10.2 "	5.8 5.8	4.6 5.3	" "	2.8 3.5	4.7 3.3	—1.3 "	3.4 2.0	10.2	—6.1	4.1
5	---	---	---	11.7 12.9	—6.4 "	5.3 6.5	16.2 16.6	" "	5.7 6.4	5.6	"	3.8	3.7	—1.4	2.3	9.5 9.4	—6.2 —6.3	3.3 3.1
6	---	---	---	14.2	"	7.8	17.0	—10.3	6.7	6.4	"	4.6	4.1	—1.5	2.6	9.4 9.5	—6.4 —6.5	3.0 3.0
7	---	---	---	15.1 15.2	—6.5 "	8.6 8.7	18.0	"	7.7	7.7	"	5.9	5.4	—1.6	3.8	9.8	—6.6	3.2
8	13.5	—5.8	7.7	15.7 15.6	—6.6 "	9.1 9.0	18.9 19.3	—10.4 "	8.5 8.9	9.6	—1.9	7.7	8.0	—1.7	6.3	11.5	—6.7	4.8
9	12.5	"	6.7	15.3 15.1	—6.7 "	8.6 7.4	19.3 19.3	—10.5 "	8.8 8.8	10.8 11.3	" "	8.9 9.4	10.2	—1.8	8.4	15.0	—6.8	8.2
10	11.2	—5.9	5.3	15.0	—6.8	8.2	19.0	—10.6	8.4	11.6 11.8	" "	9.7 9.7	11.0 12.3	—1.9 "	10.0 10.4	16.3	—6.9	9.4
11	10.4 10.5	" "	4.5 4.6	14.6	—6.9	7.7	18.5	—10.7	7.8	11.6 11.6	" "	9.7 9.7	12.4 12.4	—2.0 "	10.4 10.4	17.8	—7.0	10.8
Midn't	10.2	"	4.3	13.8	—7.0	6.8	16.4	—10.8	5.6	10.6	—2.0	8.6	12.3	—2.0	10.3	18.4	—7.1	11.3

Feb. 21. Readings irregular, tide-gauge out of order at 8 A. M., repaired at noon. Sounding 7 fath., 1 ft., 0 in.
Feb. 22 and 23. No sounding.

                          Fath. Feet. Inch.
Feb. 24. Sounding at noon 6   5   0   Corrections derived from curves.
" 25.   "   "   7   0   6   Ebb tide at 11 A. M. Corrections derived from curves.
" 26.   "   "   7   5   6   Correction derived from the means by sounding and curves.

Series II.—Tidal Observations from January 28 to April 7, 1854.

Hourly observations on the pulley-gauge. Adopted reading of mean level 7.0, expressed in units of the scale. Increasing numbers indicate rise of water.

Mean solar hour.	February, 1854.						March, 1854.								
	27th.	Red. to level.	Ref. obs.	28th.	Red. to level.	Ref. obs.	1st.	Red. to level.	Ref. obs.	2d.	Red. to level.	Ref. obs.	3d.	Red. to level.	Ref. obs.	4th.	Red. to level.	Ref. obs.
1	18.3	—7.3	11.0	15.0 14.8	—4.9 "	10.1 9.9	15.5 16.3	—5.0 "	10.5 11.3	16.2 17.0 17.0	—5.0 " "	11.2 12.0 12.0	14.5 15.5 16.0	—5.4 " "	9.1 10.1 10.6	13.1	—5.1	8.0
2	17.5	—7.4	10.1	12.8	"	7.9	16.3	"	11.3	17.0 16.5	" "	12.0 11.5	16.0 16.0	" "	10.6 10.6	15.4 15.9	—5.0 "	10.4 10.9
3	17.8	—7.5	10.3	11.0	"	6.1	13.9	"	8.9	16.0	"	11.0	16.0 15.7	" "	10.6 10.3	16.0 16.0	—4.9 "	11.1 11.1
4	19.3 5.3	? —4.8	--- 0.5	7.8	"	2.9	12.0	"	7.0	15.1	"	10.1	14.8	—5.5	9.3	15.8	"	10.9
5	4.4 4.4	" "	—0.4 —0.4	5.5 5.5	" "	0.6 0.6	10.5	"	5.5	12.8	"	7.8	13.6	"	8.1	14.2	—4.8	9.4
6	5.0	"	0.2	5.0 5.0	" "	0.1 0.1	7.2 5.5	" "	2.2 0.5	11.1	"	6.1	12.4	"	6.9	12.1	"	7.3
7	5.8	"	1.0	5.0 6.2	" "	0.1 1.3	5.5 5.5	" "	0.5 0.5	9.0 8.8	" "	4.0 3.8	11.1	—5.6	5.5 9.1	9.7	—4.7	5.0 4.4
8	8.3	"	3.5	7.0	" "	2.1	5.5 5.7	" "	0.5 0.7	7.5	"	2.5	9.4 8.1	" "	3.8 2.5	8.6 8.7	" "	3.9 4.0
9	11.1	"	6.3	10.0	"	5.1	7.0	"	2.0	8.2	"	3.2	9.0	"	3.4	9.0	"	4.3
10	14.4	"	9.6	12.9	"	8.0	9.5	"	4.5	10.0	"	5.0	9.7	—5.7	4.0 8.7	8.4	"	3.7 4.0
11	16.3 16.9	"	11.5 12.1	16.0	"	11.1	13.5	"	8.5	12.8	"	7.8	11.2	"	5.5	9.7	"	5.0
Noon	17.6 16.0	—4.9	12.7 11.1	17.7 17.9	—4.9 "	12.8 13.0	16.1	—5.0	11.1	15.6	—5.0	10.6	13.3	—5.8	7.5	12.1	—4.6	7.5
1	15.3	"	10.4	17.9 17.2	" "	13.0 12.3	17.6 17.9	" "	12.6 12.9	17.9 18.0	" "	12.9 13.0	15.7	"	9.9	13.5	"	8.9
2	14.1	"	9.2	15.9	"	11.0	17.9 17.2	" "	12.9 12.2	18.0	"	13.0	16.0 16.6	—5.7 "	10.3 10.9	15.3 15.5	" "	10.7 10.9
3	12.8	"	7.9	13.4	"	8.5	16.7	"	11.7	17.2	—5.1	12.1	16.3 16.3	" "	10.6 10.6	15.6 15.2	—4.5 "	11.1 10.7
4	8.7	"	3.8	9.7	"	4.8	13.9	"	8.9	15.8	"	10.7	15.3	"	9.6	15.0	"	10.5
5	4.7 4.5	" "	—0.2 —0.4	7.8	"	2.9	12.4	"	7.4	14.4	—5.2	9.2	14.9	—5.6	9.3	12.9	—4.4	8.5
6	4.0 4.0	" "	—0.9 —0.9	6.2 6.0	" "	1.3 1.1	11.7	"	6.7	13.4	"	8.2	14.2	"	8.6	11.2	"	6.8
7	4.7	"	—0.2	6.0 5.0	" "	1.1 1.1	11.1	"	6.1	12.0 11.3	" "	6.8 6.1	13.5	—5.5	8.0	9.9	"	5.5
8	6.5	"	1.6	6.0 6.4	" "	1.1 1.5	9.4 8.5	" "	4.4 3.5	10.0 7.7	—5.3 "	4.7 2.4	12.1 8.6	" "	6.6 3.1	8.2 6.4	—4.3 "	3.9 2.1
9	8.9	"	4.0	8.0	"	3.1	6.3 7.2	" "	1.3 2.2	7.5 7.2	" "	2.2 1.9	6.9 6.9	—5.4 "	1.5 1.5	5.2 5.3	—4.2 "	1.0 1.1
10	11.6	"	6.7	10.4	"	5.5	8.9	"	3.9	8.5	"	3.2	7.2	"	1.8	5.3 6.0	" —4.1	1.1 1.9
11	13.8 14.7	" "	8.9 9.8	15.8 17.0	" "	10.9 12.1	11.8	"	6.8	11.2	"	5.9	9.2	—5.3	3.9	6.4	"	2.3
Mid'n't	15.2	"	10.3	17.3	"	12.4	14.6	"	9.6	13.9	—5.4	8.5	11.4	—5.2	6.2	8.6	—4.0	4.6

                        Fath. Feet Inch.
Feb. 27. Sounding at noon 7 5 6   Mean correction by sounding and curves adopted, the latter
"   28. No sounding.   [showing weight 2.
March 1. No sounding.
"   2. Sounding at noon 7 5 6   Corrected reading by sounding 11.0, by curves 10.2, mean 10.6.
"   3.   "   "   7 1 6   Corrected reading by sounding 7.0, by curves 8.0, mean 7.5.
"   4. No sounding.

RECORD AND REDUCTION OF THE TIDES.

SERIES II.—TIDAL OBSERVATIONS FROM JANUARY 28 TO APRIL 7, 1854.

Hourly observations on the pulley-gauge. Adopted reading of mean level 7.0, expressed in units of the scale. Increasing numbers indicate rise of water.

March, 1854.

Mean solar hour.	5th.	Red. to level.	Ref. obs.	6th.	Red. to level.	Ref. obs.	7th.	Red. to level.	Ref. obs.	10th.	Red. to level.	Ref. obs.	11th.	Red. to level.	Ref. obs.	15th.	Red. to level.	Ref. obs.
1	10.5	—3.8	6.7	10.7	—3.5	7.2	9.7	—4.1	5.6	9.6 9.1	—4.7 "	4.9 4.4	15.1 14.3	—6.9 "	8.2 7.4	17.6 17.5	—6.4 —6.5	11.2 11.0
2	11.8	"	8.0	12.1	"	8.6	11.0	"	6.9	8.9 8.9	—4.8 "	4.1 4.1	13.2	—7.0	6.2	15.7	—6.7	9.0
3	12.7 12.8	—3.7 "	9.0 9.1	13.2	—3.6	9.6	12.1	"	8.0	8.9 9.0	—4.9 "	4.0 4.1	12.5 12.4	"	5.5 5.4	15.2	—6.9	8.3
4	12.8 11.9	—3.6 "	9.2 8.3	14.0 14.1	"	10.4 10.5	12.8 12.9	"	8.7 8.8	9.0 9.0	—5.0 "	4.0 4.0	12.4 12.4	"	5.4 5.4	14.0	—7.1	6.9
5	11.7 11.2	—3.5 "	8.2 7.7	14.1 14.0	—3.7 "	10.4 10.3	12.9 12.9	"	8.8 8.8	9.0 9.6	—5.1 —5.2	3.9 4.4	12.4 12.4	"	5.4 5.4	11.5	—7.3	4.2
6	10.9	—3.4	7.5	13.9	"	10.2 12.9	12.9	—4.2	8.7 8.7	10.0	—5.3	4.7	12.4 12.5	"	5.4 5.5	10.2 10.0	—7.5 —7.7	2.7 2.3
7	9.7	"	6.3	12.9	—3.8	9.1	12.9 12.9	"	8.7 8.7	11.3	—5.4	5.9	12.8	"	5.8	10.7	—7.9	2.8
8	8.7 7.4	—3.3 "	5.4 4.1	12.2	"	8.4	12.9 12.7	"	8.7 8.5	13.1	—5.5	7.6	14.2	"	7.2	12.4	—8.1	4.3
9	6.3 6.6	—3.2 "	3.1 3.4	10.1 10.0	—3.9 "	6.2 6.1	12.1	"	7.9	15.2 15.5	—5.6 —5.6	9.6 9.9	16.0	"	9.0	17.0	—8.3	8.7
10	6.6 6.7	—3.1 "	3.5 3.6	10.0 10.0	"	6.1 6.1	10.5 10.5	"	6.3 6.3	15.5 15.5	—5.7 "	9.8 9.8	17.2 17.5	"	10.2 10.5	19.2	—8.5	10.7
11	7.3	"	4.2	10.0 10.4	"	6.1 6.5	10.0 10.2	"	5.8 6.0	15.3	—5.8	9.5	17.6	"	10.6	20.7 21.5	—8.7 —8.8	12.0 12.7
Noon	9.5	—3.0	6.5	10.4	—4.0	6.4	10.2	—4.3	5.9	15.0	—5.9	9.1	---			18.0	—8.9	9.1
1	10.5	"	7.5	11.3	"	7.3	10.5	"	6.2	14.0	—6.0	8.0				18.3	—9.0	9.3
2	11.8 12.3	—3.1 "	8.7 9.2	12.0	"	8.0				11.9	—6.1	5.8				17.3	—9.1	8.2
3	13.2 13.2	"	10.1 10.1	12.6	"	8.6				11.5 11.4	—6.2 "	5.3 5.2				16.0	—9.2	6.8
4	13.2 12.7	—3.2 "	10.0 9.5	12.9	"	8.9				11.3 11.3	" —6.3	5.1 5.0				13.7	"	4.5
5	11.9	"	8.7	13.3 13.6	"	9.3 9.6				11.3 11.6	" "	5.0 5.3				12.9	—9.3	3.6
6	10.4	—3.3	7.1	13.3	"	9.3				11.8	—6.4	5.4				12.6 12.4	—9.4 "	3.2 3.0
7	9.8	"	6.5	12.7	"	8.7				12.3	—6.5	5.8				12.3 12.3	—9.5 "	2.8 2.8
8	8.8	"	5.5	12.2	"	8.2				12.9	"	6.4				12.3 12.3	—9.6 "	2.7 2.7
9	7.7 7.4	—3.4 "	4.3 4.0	9.5	"	5.5				13.5	—6.6	6.9				14.0	—9.7	4.3
10	7.3 7.5	" "	3.9 4.1	8.2 8.1	"	4.2 4.1				15.0 15.5	—6.7 "	8.3 8.8				16.5	—9.8	6.7
11	7.8	"	4.4	8.1 8.4	"	4.1 4.4				15.5 15.5	" "	8.8 8.8				19.1	—9.9	9.2
Midn't	9.2	—3.5	5.7	8.6	—4.1	4.5				15.5	—6.8	8.7				20.5	—10.0	10.5

March 5. Soundings at noon 6 fath., 3 feet, 6 inches. Correction from curves.
   " 7. Tide register broke at 9 A. M., was repaired immediately. No sounding.
March 6. No sounding.
March 8. No sounding.

                                Fath. Feet Inch.

   " 9. Sounding at noon 6 4 0
   " 10.   "   "   7 0 0
   " 11.   "   "   7 — —   March 12. No sounding and but a few observations taken.
   " 14.   "   "   7 5 6   " 13. But few observations taken.
   " 15.   "   "   8 — —   The corrections after March 7 are derived from curves.

Series II.—Tidal Observations from January 28 to April 7, 1854.

Hourly observations on the pulley-gauge. Adopted reading of mean level 7.0, expressed in units of the scale. Increasing numbers indicate rise of water.

March, 1854.

Mean solar hour.	16th.	Red. to level.	Ref. obs.	17th.	Red. to level.	Ref. obs.	18th.	Red. to level.	Ref. obs.	19th.	Red. to level.	Ref. obs.	20th.	Red. to level.	Ref. obs.	23d.	Red. to level.	Ref. obs.
1	20.5 20.5 20.5	—10.0 " "	10.5 10.5 10.5	21.4 21.6	—9.6 "	11.8 12.0	19.5 20.3	—8.8 "	10.7 11.5	21.4	—10.3	11.1	19.8 20.9	—11.0 "	8.8 9.9			Irregular
2	20.0	"	10.0	21.6 21.0	" "	12.0 11.4	20.9 20.8	" "	12.0 12.0	22.2 22.7	—10.4 "	11.8 12.3	21.3 21.5	" "	10.3 10.5			
3	17.5		7.5	19.8	—9.5	10.3	19.5	"	10.7	22.7 22.7	—10.5 "	12.2 12.2	21.7 22.4	" "	10.7 11.4			
4	15.4	"	5.4	17.5	"	8.0	17.7	—8.7	9.0	22.2	—10.6	11.6	22.4 22.4	" "	11.4 12.4	10.4	—4.0	6.4
5	15.1	—10.1	5.0	16.7	—9.4	7.3	15.9	"	7.2	19.2	"	8.6	22.2	"	11.2	10.5	—4.1	6.4
6	14.8	"	4.7	14.8	"	5.4	12.4	"	3.7	17.0	—10.7	6.3	19.3	"	8.3	11.5 12.3	—4.2 "	7.3 8.1
7	14.3	"	4.2	12.7	"	3.3	11.2	"	2.5	15.0 14.1	" "	4.3 3.4	19.0	"	8.0	13.2 13.7	" "	9.0 9.5
8	13.9 13.3	" "	3.8 3.2	10.3 10.0	" —9.3	0.9 0.7	10.7 10.5	" "	2.0 1.8	13.4 14.1	" "	2.7 3.4				13.2	—4.3	8.9
9	15.0	"	5.5	11.2	"	1.9	10.3 10.3	" "	1.6 1.6	14.7	—10.8	3.9				12.6	—4.4	8.2
10	16.5	"	6.4	16.0	"	6.7	10.3	"	1.6	16.2	"	5.4				12.0	—4.5	7.5
11	18.3	"	8.2	19.6	—9.2	10.4	13.3	"	4.6	17.2	"	6.4				11.0	—4.7	6.3
Noon	18.3 18.4	—10.2 "	8.1 8.2	17.0	?	---	14.6	—8.6	6.0	19.3	"	8.5				11.2	—4.8	6.4
1	18.6 18.5	" "	8.4 8.3	19.6	—9.2	10.4	16.9	"	8.3	20.3	"	9.5		Observations irregular.		11.1 11.1	—5.0 "	6.1 6.1
2	18.5	—10.1	8.4	19.5	"	10.3	18.2 18.1	—8.7 "	9.5 9.4	21.5 21.6	" —10.9	10.7 10.7				10.4 10.5	—5.1 "	5.3 5.4
3	15.0	"	4.9	18.6	—9.1	9.5	18.1	"	9.4	21.6 21.6	" "	10.7 10.7				11.1	—5.2	5.9
4	13.5 13.2	—10.0 "	3.5 3.2	17.2	"	8.1	16.0	—8.8	7.2	21.4	"	10.5				11.8	—5.4	6.4
5	13.0 13.0	" "	3.0 3.0	16.8	"	7.7	14.6	—8.9	5.7	20.2	"	9.3				12.2	—5.5	6.7
6	13.0 13.0	—9.9 "	3.1 3.1	15.7	—9.0	6.7	14.1	—9.0	5.1	17.2	"	6.3				13.4	—5.6	7.8
7	13.0 13.0	" "	3.1 3.1	14.3	"	5.3	13.1 12.6	—9.1 —9.2	4.0 3.4	13.9 13.7	" "	3.0 2.8				13.9	—5.7	8.2
8	13.0 13.3	" "	3.1 3.4	12.1 10.2	" "	3.1 1.2	12.4 12.5	—9.3 —9.4	3.1 3.1	13.5 13.5	" "	2.6 2.6				14.2 14.3	—5.8 "	8.4 8.5
9	14.8	—9.8	5.0	10.4 11.3	" "	1.4 2.3	13.0	—9.6	3.4	13.5 13.5	" "	2.6 2.6				14.0	"	8.2
10	15.8	"	6.0	13.2	—8.9	4.3	14.2	—9.8	4.4	14.0	"	3.1				13.8	—5.9	7.9
11	18.3	"	8.5	13.6	"	4.7	17.2	—10.0	7.2	18.4	"	7.5				12.2	"	6.3
Midn't	20.0	—9.7	10.3	13.9	"	5.0	19.2	—10.2	9.0	20.9	—11.0	9.9				12.0	—6.0	6.0

	Fath.	Feet.	Inch.	
March 16. Sounding at noon	8	—	—	⎫
" 17. " "	7	5	6	⎪
" 18. " "	7	3	0	⎬ Corrections from curves.
" 19. No sounding.				⎪
" 20. " "				⎪
" 21. Sounding at noon	6	1	0	⎪
" 22. " "	6	0	0	⎪
" 23. No sounding.				⎭

RECORD AND REDUCTION OF THE TIDES. 25

SERIES II.—TIDAL OBSERVATIONS FROM JANUARY 28 TO APRIL 7, 1854.

Hourly observations on the pulley-gauge. Adopted reading of mean level 7.0, expressed in units of the scale. Increasing numbers indicate rise of water.

March, 1854.

Mean solar hour.	24th.	Red. to level.	Ref. obs.	25th.	Red. to level.	Ref. obs.	26th.	Red. to level.	Ref. obs.	27th.	Red. to level.	Ref. obs.	28th.	Red. to level.	Ref. obs.	29th.	Red. to level.	Ref. obs.
1	11.3	—6.0	5.3	8.5	—0.1	8.4	9.9	—2.9	7.0	10.2	—1.1	9.1	15.0	—4.7	10.3	15.5 15.3 14.0	—3.5 —3.3 "	12.0 12.0 10.7
2	10.9 10.9 10.9	" —6.1 "	4.9 4.8 4.8	6.5 5.5	—0.2 "	6.3 5.3	8.3	—3.0	5.3	8.2	—1.3	6.9	12.4	—4.9	7.5	13.3	—3.2	10.1
3	11.2	—6.2	5.0	5.5 5.5	—0.3 "	5.2 5.2	7.5 7.3	—3.2 —3.3	4.3 4.0	6.0	—1.4	4.6	9.7	—5.1	4.6	11.3	—3.1	8.2
4	12.1	"	5.9	5.5 5.6	—0.4 "	5.1 5.2	7.3 7.3	—3.4 —3.5	3.9 3.8	4.1 3.9	—1.5 —1.6	2.6 2.3	9.0	—5.2	3.8	9.0	—3.0	6.0
5	12.9	—6.3	6.6	5.7	—0.5	5.2	7.6	—3.6	4.0	5.9 4.2	—1.7 —1.8	2.2 2.4	8.0 7.7	—5.3 —5.4	2.7 2.3	7.5 7.1	" "	4.5 4.1
6	13.8	"	7.5	8.2	—0.7	7.5	9.2	—3.8	5.4	4.5	—1.9	2.6	7.5 8.2	—5.5 —5.6	2.6 2.6	6.5 6.5	" "	3.5 3.5
7	15.0 16.0	—6.4 "	8.6 9.6	9.2	—0.8	8.4	11.8	—4.0	7.8	5.3	—2.0	3.3	8.7	—5.7	3.0	6.5 7.2	" "	3.5 4.2
8	16.7 16.4	—6.5 "	10.2 9.9	9.9	—1.0	8.9	13.5	—4.2	9.3	8.0	—2.1	5.9	11.2	—5.8	5.4	8.6	"	5.6
9	16.0	—6.6	9.4	11.2 11.5	—1.1 —1.2	10.1 10.3	16.0 16.2	—4.4 —4.5	11.6 11.7	12.1	—2.2	9.9	16.4	"	10.6	12.8	"	9.8
10	15.5	—6.7	8.8	11.5 11.5	—1.3 "	10.2 10.2	16.2 15.5	—4.7 —4.9	11.5 10.6	13.7 14.0	—2.4 —2.5	11.3 11.5	16.9 17.2	" "	11.1 11.4	14.9	"	11.9
11	15.0	—6.8	8.2	11.5 10.5	—1.4 "	10.1 9.1	14.0	—5.2	8.8	14.0 14.0	—2.1 —2.7	11.9 11.3	17.7 17.8	" "	11.9 12.0	16.5	"	13.5
Noon	8.7 ---	—0.7	8.0	9.5	—1.5	8.0	12.0 ---	—5.5	6.5	13.9	—2.8	11.1	16.0	—5.9	10.1 17.5	17.2 17.5	—3.0 "	14.2 14.5
1	7.7	—0.5	7.2	8.7	—1.6	7.1	10.3	?	...	12.6	—2.9	9.7	16.3	"	10.4	17.5 17.4	" "	14.5 14.4
2	5.8 5.6	—0.3	5.5 3.3	7.7	—1.7	6.0	8.7	"	...	9.4	—3.1	6.3	16.3	—5.8	10.5	16.9	"	13.9
3	5.6 5.6	—0.1 "	5.5 5.5	6.6	—1.9	4.7	6.2	"	...	6.6	—3.2	3.4	16.6	—5.6	11.0	11.0	"	8.0
4	5.6	0.0	5.6	6.3	—2.0	4.3	3.7 3.4	0.0 "	3.7 3.4	6.5 4.9	—3.4 "	3.1 1.5	12.8	—5.3	7.5	7.7	"	4.7
5	6.0	"	6.0	6.6 7.3	—2.1 "	4.5 5.2	3.5	—0.1	3.4	5.1 4.9	—3.5 —3.6	1.6 1.3	11.5	—5.1	6.4	5.0	"	2.0
6	6.8	"	6.8	7.8	—2.2	5.6	3.9	—0.3	3.6	5.2	—3.7	1.5	10.2 3.5 5.2	—4.9 —4.8 "	5.3 —1.3 0.5	2.9 2.8	" "	—0.1 —0.2
7	7.6	"	7.6	9.3	—2.3	7.0	4.6	—0.4	4.2	6.9	—3.9	3.0	3.8	—4.7	—0.9	3.2	"	0.3
8	7.7	"	7.7	10.9	—2.4	8.5	6.9	—0.5	6.4	8.6	—4.1	4.5	6.2	—4.6	1.6	5.2	"	2.2
9	8.3	"	8.3	11.4 12.0	—2.5 "	8.9 9.5	9.9	—0.6	9.3	12.3	—4.2	8.1	11.6	—4.4	7.2	7.2	"	4.2
10	9.1 9.2	" "	9.1 9.2	12.2 12.0	—2.6 "	9.6 9.4	11.2	—0.8	10.4	13.9 14.4	—4.3 "	9.6 10.1	12.6	—4.2	8.4	10.5	"	7.5
11	9.2 8.7	" "	9.2 8.7	12.0	—2.7	9.3	11.9 12.3	—0.9 "	11.0 11.4	14.6 15.3	—4.4 "	10.2 10.9	14.6	—4.0	10.6	13.7	"	10.7
Midn't	8.4	"	8.4	11.5	—2.8	8.7	11.4	—1.0	10.4	15.2	—4.5	10.7	15.5	—3.7	11.8	15.2	"	12.2

```
                            Fath.  Feet.  Inch.
March 24.  Sounding at noon   6      2      6
    "  25.       "       "    6      3      6
    "  26.       "       "    7      0      0
    "  27.  No sounding.
    "  28.  Sounding at noon  7      4      0   Correction by sounding —6.6, by curve —5.2, mean —5.9
    "  29.       "       "    7      5      6
```

SERIES II.—TIDAL OBSERVATIONS FROM JANUARY 28 TO APRIL 7, 1854.

Hourly observations on the pulley-gange. Adopted reading of mean level 7.0, expressed in units of the scale. Increasing numbers indicate rise of water.

March, 1854. | April, 1854.

Mean solar hour.	30th.	Red. to level.	Ref. obs.	31st.	Red. to level.	Ref. obs.	4th.	Red. to level.	Ref. obs.	5th.	Red. to level.	Ref. obs.	6th.	Red. to level.	Ref. obs.	7th.	Red. to level.	Ref. obs.
1	15.3 15.3	—3.0 "	12.3 12.3	13.8	—3.0	10.8	...			8.3 9.6	—3.0 "	5.3 6.6	7.5 7.8	—1.7 "	5.8 6.1	7.0 7.5	—1.5 "	5.5 6.0
2	15.4 15.5 15.3	" " "	12.4 12.5 12.3	14.5 14.5 13.2	" " "	11.5 11.5 10.2	...			10.5	—2.9	7.6	8.4	—1.6	6.8	7.5	"	6.0
3	14.6	"	11.6	12.0	"	9.0	...			11.2	—2.8	8.4	9.6	"	8.0	7.5	"	6.0
4	13.6	"	10.6	9.7	"	6.7	...			11.8	"	9.0	10.0	"	8.4	8.0	"	6.5
5	11.8	"	8.8	7.0	"	4.0	...			12.2 12.4	—2.7 "	9.5 9.7	10.4 10.4	" "	8.8 8.8	8.6	"	7.1
6	9.4	"	6.4	5.6	"	2.8	...			12.4	"	9.7	10.4	"	8.8	9.8 9.9	" "	8.3 8.4
7	6.9	"	3.9	5.0 4.7	" "	2.0 1.7	...			11.2	—2.6	8.6	10.4 10.4	" "	8.8 8.8	10.0 10.0	" "	8.5 8.5
8	6.3	"	3.3	4.5 5.0	" "	1.5 2.0	...			9.2	—2.5	6.7	10.4 10.2	" "	8.6 8.6	10.0 10.0	" "	8.5 8.5
9	7.2	"	4.2	6.0	"	3.0	...			8.1	"	5.6	10.0	"	8.4	10.0 10.0	" "	8.5 8.5
10	10.1	"	7.1	8.0	"	5.0	...			7.4 7.4	—2.4 "	5.0 5.0	9.1	"	7.5	9.5	"	8.0
11	12.9 15.2	" "	9.9 12.2	10.1		7.1	...			7.4 7.4	" "	5.0 5.0	8.5	"	6.9	8.8	"	7.3
Noon	16.5 16.1	" "	13.5 13.1	11.9		8.9	9.0	—4.0	5.0	8.0	—2.3	5.7	8.0	—1.5	6.5	8.5	—1.5	7.0
1	15.3	"	12.3	14.7 16.0	" "	11.7 13.0	9.7	"	5.7	8.6	"	6.3	7.5	"	6.0	7.0 7.0	" "	5.5 5.5
2	13.3	"	10.3	15.8 16.0	" "	12.8 13.0	11.0	—3.9	7.1	8.0	—2.2	6.7	7.5 7.5	" "	6.0 6.0	7.0 7.0	" "	5.5 5.5
3	12.3	"	9.3	15.2	"	12.2	11.3 12.3	" —3.8	7.4 8.5	8.9	—2.1	6.8	7.5	"	6.0	7.0 7.0	" "	5.5 5.5
4	8.1	"	5.1	14.7	"	11.7	13.1 12.9	" —3.7	9.3 9.2	8.0	"	6.8	7.5	"	6.0	7.0	"	5.5
5	7.0	"	4.0	12.9	"	9.9	11.7	"	8.0	8.6 8.9	—2.0 "	6.6 6.9	7.9	"	6.4	7.5	"	6.0
6	5.5	"	2.5	10.8	"	7.8	9.9	—3.6	6.3	8.9 8.7	" —1.9	6.9 6.8	8.0	"	6.5	8.1	"	6.6
7	4.4 4.3	" "	1.4 1.3	11.5	"	8.5	9.0	"	5.4	8.7	"	6.8	8.0	"	6.5	8.7	"	7.2
8	4.3 4.4	" "	1.3 1.4	8.4 7.0	—2.9 "	5.5 4.1	8.0	—3.5	4.5	8.1	"	6.2	8.0	"	6.5	9.0	"	7.5
9	6.8	"	3.8	6.1 7.6	—2.8 "	3.3 4.8	8.2 7.8	—3.4 "	4.8 4.4	8.0	—1.8	6.2	8.2	"	6.7	9.0 9.0	" "	7.5 7.5
10	10.3	"	7.3	8.8	—2.7	6.1	7.5 7.5	—3.3 "	4.2 4.2	7.5 7.5	" "	5.7 5.7	7.6	"	6.3	9.0 9.0	" "	7.5 7.5
11	12.2	"	9.2	11.4 12.2	—2.6 "	8.8 9.6	7.5 7.7	—3.2 "	4.3 4.5	7.5 7.5	" "	5.7 5.7	7.3 7.0	" "	5.8 5.5	9.0	"	7.5
Midn'h	13.1	"	10.1	13.0 12.4	—2.5 "	10.5 9.9	8.3	—3.1	5.2	7.5	—1.7	5.8	7.0	"	5.5	9.0	"	7.5

```
                          Fath.  Feet  Inch.
March 30, Sounding at noon   7     4     6     Correction by curves preferred.
  "    31, No sounding.
April 1, 2,   "      "
  "    4, Sounding at noon   6     2     0  ⎫
  "    5,    "       "       6     2     0  ⎬ Corrections derived from curves, readings (heights) not
  "    6,    "       "       6     2     6  ⎪ reliable, see preceding note of April 14.
  "    7,    "       "       6     2     6  ⎭
```

RECORD AND REDUCTION OF THE TIDES. 27

SERIES III.—TIDAL OBSERVATIONS FROM APRIL 20 TO AUGUST 3, 1854.

Hourly observations on the pulley-gauge. Adopted reading of mean level 7.0, expressed in units of the scale. Increasing numbers indicate rise of water.

April, 1854.

Mean solar hour.	20th.	Red. to level.	Ref. obs.	21st.	Red. to level.	Ref. obs.	22d.	Red. to level.	Ref. obs.	23d.	Red. to level.	Ref. obs.	24th.	Red. to level.	Ref. obs.	25th.	Red. to level.	Ref. obs.
1				6.7	—2.2	4.5	7.8	—3.5	4.3									
				7.3	"	5.1	7.5	"	4.0	10.0	—4.2	5.8	9.4	—4.0	5.4	11.1	—3.5	7.6
							7.5	"	4.0									
2				7.9	—2.3	5.6	7.7	—3.6	4.1	9.9	"	5.7	7.5	"	3.5	8.8	"	5.3
										8.5	"	4.3						
3				9.0	—2.4	6.6	7.9	"	4.3	8.3	"	4.1	6.7	"	2.7	7.5	"	4.0
										8.5	"	4.3	6.5	"	2.5			
4	Irregular.			10.2	"	7.8	8.9	—3.7	5.2	8.7	"	4.5	6.5	—3.9	2.6	6.4	—3.4	3.0
													6.7	"	2.8	5.8	"	2.4
5				11.4	—2.5	8.9	9.7	"	6.0	10.2	"	6.0	7.1	"	3.2	5.0	"	1.6
																6.0	"	2.6
6				12.1	"	9.6	12.2	—3.8	8.4	11.8	"	7.6	8.8	"	4.9	7.0	"	3.6
7				12.5	—2.6	9.9	13.3	"	9.5	13.2	"	9.0	10.5	"	6.6	8.5	"	5.1
				12.6	"	10.0	13.7	"	9.9	14.0	"	9.8						
8				12.5	—2.7	9.8	14.3	—3.9	10.4	14.9	"	10.7	12.9	"	9.0	10.9	—3.3	7.6
							14.0	"	10.1	14.3	"	10.7	15.0	"	11.1			
9	11.3	—1.0	10.3	11.0	—2.8	8.2	13.7	"	9.8	13.0	"	8.8	15.0	"	11.1	13.4	"	10.1
																14.5	"	11.2
10	8.3	—1.1	7.2	9.9	"	7.1	12.3	"	8.4	8.1	?	---	14.9	"	11.0	15.5	"	12.2
																15.3	"	12.0
11	6.7	—1.2	5.5	8.5	—2.9	5.6	11.3	"	7.4	10.7	"	---	14.6	"	10.7	14.1	"	10.8
	6.4	"	5.2															
Noon	6.4	—1.3	5.1	7.9	—3.0	4.9	10.7	—4.0	6.7	11.5	—4.3	7.2	12.9	—3.8	5.1	12.5	—3.2	9.3
	6.4	"	5.1	7.4	"	4.4												
1	6.7	"	5.4	7.0	"	4.0	8.7	"	4.7	9.5	"	5.2	10.6	"	6.8	10.3	"	7.1
				7.0	"	4.0												
2	7.1	—1.4	5.7	7.0	—3.1	3.9	8.3	"	4.3	8.3	"	4.0	9.0	"	5.2	8.3	"	5.1
				7.4	"	4.3	8.0	"	4.0									
3	7.9	"	6.5	8.0	"	4.9	8.0	"	4.0	7.2	—4.2	3.0	7.4	"	3.6	6.9	"	3.7
							8.0	"	4.0	6.6	"	2.4						
4	9.0	—1.5	7.5	8.8	—3.2	5.6	8.0	"	4.0	6.1	"	1.9	6.1	—3.7	2.4	5.2	"	2.0
							8.3	"	4.3	6.0	"	1.8	5.4	"	1.7	5.0	"	1.8
5	9.5	"	8.0	8.5	"	5.3	9.0	—4.1	4.9	6.5	—4.1	2.4	5.7	"	2.0	5.0	"	1.8
	9.7	"	8.2													5.0	"	1.8
6	9.7	—1.6	8.1	9.2	—3.3	5.9	10.8	"	6.7	7.5	"	3.4	8.1	"	4.4	5.2	"	2.0
	9.7	"	8.1	10.1	"	6.8												
7	9.5	"	7.9	10.5	"	7.2	11.5	"	7.4	9.3	"	5.2	9.0	"	5.3	7.5	"	4.3
				10.5	"	7.2												
8	8.8	—1.7	7.1	10.5	"	7.2	12.5	"	8.4	10.8	"	6.7	10.4	—3.6	6.8	9.7	—3.1	6.6
				10.5	"	7.2												
9	7.9	—1.8	6.1	10.5	—3.4	7.1	13.0	"	8.9	12.0	"	7.9	13.5	"	9.9	---		
				10.0	"	6.6	13.9	"	9.8	12.8	"	8.7	13.5	"	9.9			
10	7.0	—1.9	5.1	9.1	"	5.7	13.9	"	9.8	13.6	—4.0	9.6	13.5	"	9.9	---		
							13.0	—4.2	8.8	13.6	"	9.6	12.5	"	8.9			
11	6.9	—2.0	4.9	8.4	"	5.0	12.0	"	7.8	13.0	"	9.0	12.0	"	8.4	---		
Midn't	6.9	—2.1	4.8	7.9	—3.5	4.4	11.3	"	7.1	9.0	"	5.0	11.5	—3.5	8.0	---		

April 20. No sounding.

                       Fath. Feet. Inch.

" 21. Sounding at noon 6 1 0 ⎫
" 22.   "   " 6 3 0 ⎪
" 23.   "   " 6 3 6 ⎬ The corrections were deduced from the curves.
" 24.   "   " 6 4 5 ⎪
" 25.   "   " 7 0 6 ⎭

Series III.—Tidal Observations from April 20 to August 3, 1854.

Hourly observations on the pulley-gauge. Adopted reading of mean level 7.0, expressed in units of the scale. Increasing numbers indicate rise of water.

April, 1854. | May, 1854.

Mean solar hour.	26th.	Red. to level.	Ref. obs.	27th.	Red. to level.	Ref. obs.	28th.	Red. to level.	Ref. obs.	29th.	Red. to level.	Ref. obs.	30th.	Red. to level.	Ref. obs.	1st.	Red. to level.	Ref. obs.
1	10.1	—3.0	7.1	14.4 14.5 14.1	—3.0	"	11.4 11.5 11.1	—3.0	"	9.2	—3.0	6.2	16.4 16.0	—3.4	13.0 12.6	16.4 15.3	—3.6	12.8 11.7	16.2 16.5 16.5	" " "	12.3 12.6 12.6
2	9.1	"	6.1	12.8	"	"	9.8	7.8	"	4.8	14.9	"	11.5	12.8	"	9.2	15.9	—4.0	11.9
3	7.1 6.3	"	4.1 3.3	10.5	"	"	7.5	7.0	"	4.0	13.1	"	9.7	10.3	"	6.7	14.7	"	10.7
4	5.6 5.8	"	2.6 2.8	8.3	"	"	5.3	6.0 5.3	"	3.0 2.3	11.7	"	8.3	7.6	"	4.0	13.4	"	9.4
5	6.3	"	3.3	6.8	"	"	3.8	5.0 5.1	"	2.0 2.1	9.8	"	6.4	6.2 5.5	"	2.6 1.9	12.2	"	8.2
6	7.3	"	4.3	4.7 4.0 1.0	"	"	1.7	5.7	"	2.7	8.0	"	4.6	6.3	"	2.6	10.0	"	6.0
7	8.5	"	5.5	4.0 5.9	"	"	1.0 2.9	6.6	—3.1	3.5	5.9 5.5	"	2.5 2.1	7.7	"	4.1	8.9	"	4.9
8	9.4	"	6.4	8.7	"	"	5.7	7.5	"	4.4	6.2	"	2.6	9.1	"	5.5	8.2 8.0	"	4.2 4.0
9	10.6	"	7.6	10.9	"	"	7.9	8.8	"	5.7	8.3	"	4.9	10.4	"	6.5	7.9 6.4	"	3.9 4.4
10	11.4	"	8.4	12.7	"	"	9.7	10.3	"	7.2	10.9	"	7.5	11.2	"	7.6	9.7	"	5.7
11	13.9 14.4	" "	10.9 11.4	14.4 14.5	" "	" "	11.4 11.5	11.7	"	8.6	13.5 14.6	"	10.1 11.2	13.0	"	9.4	11.9	"	7.9
Noon	15.0 14.0	" "	12.0 11.0	15.4 15.2	" "	" "	12.1 12.2	13.7 14.5	—3.2	10.5 11.3	15.5 14.5	—3.5	12.0 11.0	14.7 15.0	—3.7	11.0 11.3	13.1 13.5	—4.1	9.0 9.4
1	13.1	"	10.1	15.0	"	"	12.0	14.8 14.4	" "	11.6 11.2	13.2	"	9.7	14.4	"	10.7	13.0	"	8.9
2	10.7	"	7.7	12.4	"	"	9.4	14.3	"	11.1	10.9	"	7.4	12.6	"	8.9	11.3	"	7.2
3	8.7	"	5.7	10.3	"	"	7.3	12.4	"	9.2	9.6	"	6.1	10.2	"	6.5	10.2	—4.2	6.0
4	6.1	"	3.1	7.3 6.2	" "	" "	4.3 3.2	9.2	"	6.0	7.1	"	3.6	8.0 6.3	"	4.3 2.6	9.8	"	5.6
5	4.1	"	1.1	5.2 5.9	" "	" "	2.2 2.9	6.0	"	2.8	5.3 5.0	—3.5	1.8 1.5	5.5 5.5	—3.8	1.7 1.7	9.1 8.6	"	4.9 4.4
6	3.2 3.1	" "	0.2 0.1	7.7	"	"	4.7	5.5 5.3	"	2.3 2.1	5.1	"	1.6	6.2	"	2.4	8.0 8.0	"	3.8 3.8
7	4.6	"	1.6	9.6	"	"	6.6	5.3 6.1	—3.3	2.0 2.8	7.1	"	3.6	7.6	"	3.8	8.0 8.4	—4.3	3.7 4.1
8	7.2	"	4.2	10.8	"	"	7.8	7.8	"	4.5	8.6	"	5.1	8.5	"	4.7	8.8	"	4.5
9	10.7	"	7.7	12.3	"	"	9.3	10.2	"	6.9	10.4	"	6.9	9.5	"	5.7	10.2	"	5.9
10	12.3	"	9.3	14.0	"	"	11.0	12.8	"	9.5	13.4	"	9.9	11.3	"	7.5	11.3	"	7.0
11	14.2	"	11.2	15.7	?	"	14.8	"	11.5	15.3 16.2	" "	11.8 12.7	13.9	—3.9	10.0	12.7	"	8.4
Midn't	14.3	"	11.3	16.7	"	"	15.5	—3.3	12.2	16.4	"	12.9	15.4	"	11.5	14.1	—4.4	9.7

	Fath.	Feet	Inch.	
April 26. Sounding at noon	7	3	0	Corrections deduced from the curves.
" 27. " "	7	3	2	" " " "
" 28–30. No sounding.				
May 1. Sounding at noon	7	4	0	

RECORD AND REDUCTION OF THE TIDES.

Series III.—Tidal Observations from April 20 to August 3, 1854.

Hourly observations on the pulley-gauge. Adopted reading of mean level 7.0, expressed in units of the scale. Increasing numbers indicate rise of water.

May, 1854.

Mean solar hour.	2d.	Red. to level.	Ref. obs.	3d.	Red. to level.	Ref. obs.	4th.	Red. to level.	Ref. obs.	7th.	Red. to level.	Ref. obs.	8th.	Red. to level.	Ref. obs.	9th.	Red. to level.	Ref. obs.
1	15.1	—4.4	10.7	13.2	"	8.5	13.6	—5.2	8.4	10.2 10.3 10.5	—4.2 " "	6.0 6.0 6.3	10.5 10.0 11.5	—4.8 " "	5.7 5.2 6.7	10.7 10.0	—5.5 "	5.2 4.5
2	16.7 16.8	" "	12.3 12.4	14.0	"	9.3	14.0	—5.3	8.7	10.8	"	6.6	11.5	—4.9	6.6	10.0 10.5	—5.6 "	4.4 4.9
3	16.5	"	12.1	14.2 14.3	" "	9.5 9.6	14.8	—5.4	9.4	11.3	"	7.1	11.8	"	6.9	11.3	"	5.7
4	14.2	—4.5	9.7	14.3 14.2	" "	9.6 9.5	16.1	—5.6	10.5	11.8	"	7.6	12.4	"	7.5	12.6	—5.7	6.9
5	12.5	"	8.0	14.1	"	9.4	16.2	—5.8	10.4	12.4	"	8.2	13.0	"	8.1	14.3	"	8.6
6	11.0	"	6.5	13.1	"	8.4 15.7	16.2 —6.1	—6.0	10.2 9.6	13.2	—4.3	8.9	14.1	—5.0	9.1	15.6	"	9.9
7	8.4	"	3.9	12.0	"	7.3	15.3	—6.3	9.0	13.2	"	8.9	16.2	"	11.2	16.4 16.9	—5.8 "	10.6 11.1
8	6.7 6.4	" "	2.2 1.9	10.9 10.2	" "	6.2 5.5				13.0	"	8.7	16.0 16.5	" "	11.0 11.5	17.2 17.2	—5.9 "	11.3 11.3
9	6.4	—4.6	1.8	10.0 10.3	" "	5.3 5.6				12.2	"	7.9	16.5 16.3	—5.1 "	11.4 11.2	17.2 16.6	—6.0 "	11.2 10.6
10	7.4	"	2.8	10.5 10.8	" "	5.8 6.1				11.0	"	6.7	16.0	"	10.9	16.0	—6.1	9.9
11	9.5	"	4.9	11.0	"	6.3				10.2	"	5.9	15.1	"	10.0	14.9	—6.2	8.7
Noon	11.3	—4.7	6.6	11.4	4.7	6.7				9.0	—4.4	4.6	14.1	—5.2	8.9	13.8	—6.3	7.5
1	13.1	"	8.4	12.8 13.4	" "	8.1 8.7				8.1 8.0	" "	3.7 3.6	13.0	"	7.8	?	"	---
2	14.9 15.0	" "	10.2 10.3	13.8 13.8	" "	9.1 9.1				8.0 8.0	" "	3.6 3.6	10.6	"	5.7	11.2	"	4.9
3	14.5	"	9.8	13.8 13.8	" "	9.1 9.1				8.4	"	4.0	9.0 8.4	—5.3 "	3.7 3.1	10.0 9.3	—6.2 "	3.8 3.1
4	13.6	"	8.9	13.8 13.1	" "	9.1 8.4	Readings irregular.			9.0	—4.5	4.5	8.0 8.7	" "	2.7 3.4	8.0 8.0	" "	1.8 1.8
5	12.6	"	7.9	12.7	"	8.0				9.8	"	5.3	9.0	—5.4	3.6	8.9	"	2.7
6	10.1	"	5.4	12.1	—4.8	7.3				10.3	"	5.8	10.2	"	4.8	11.2	—6.1	5.1
7	9.5	"	4.8	11.8	"	7.0				11.0	—4.6	6.4	11.6	"	6.2	13.0	"	6.9
8	9.0 9.0	" "	4.3 4.3	11.6 11.6	" "	6.7 6.7	—4.9			11.3	"	6.7	12.3	"	6.9	14.2	"	8.1
9	9.0 9.1	" "	4.3 4.4	11.4 11.6	" "	6.5 6.7				12.0 12.5	" "	7.4 7.9	12.9 13.3	" "	7.5 7.9	15.5	"	9.4
10	9.5	"	4.8	12.0	—5.0	7.0				12.5 12.5	—4.7 "	7.8 7.8	13.5 13.5	" "	8.1 8.1	14.6	"	8.5
11	10.1	"	5.4	12.7	"	7.7				12.0	"	7.3	13.3	"	7.9	14.0	"	7.9
Midn't	11.1	"	6.4	13.3	—5.1	8.2				10.5	—4.8	5.7	12.7	—5.5	7.2	13.5	—6.0	7.5

```
                        Fath. Feet. Inch.
May 2. Sounding at noon   7    1     0   Correction derived from sounding and curves.
  "  3.     "       "     7    2     0      "       "        "        "       "
  "  4.     "       "     7    1     0
  "  5.     "       "     6    5     0   From this day there is but one reading at each half hour.
  "  6. No sounding.
  "  7. Sounding at noon  6    3     0
  "  8.     "       "     7    2     0
  "  9.     "       "     7    1     6   Correction from sounding and curves.
```

Series III.—Tidal Observations from April 20 to August 3, 1854.

Hourly observations on the pulley-gauge. Adopted reading of mean level 7.0, expressed in units of the scale. Increasing numbers indicate rise of water.

May, 1854.

Mean solar hour.	10th.	Red. to level.	Ref. obs.	11th.	Red. to level.	Ref. obs.	12th.	Red. to level.	Ref. obs.	13th.	Red. to level.	Ref. obs.	14th.	Red. to level.	Ref. obs.	15th.	Red. to level.	Ref. obs.
1	12.5	—6.0	6.5	17.0 16.5	" —6.3	10.8 10.2	17.4	—6.3	11.1	17.3	—6.5	10.8	18.3	—6.7	11.6	20.6 20.6 19.7	—6.7 " "	13.9 13.9 13.0
2	9.8	"	3.8	13.7	"	7.4	16.0	"	9.7	16.1	"	9.6	16.2	"	9.5	18.8	"	12.1
3	9.5	"	3.5	11.5	"	5.2	13.7	"	7.4	15.3	"	8.8	12.3	"	5.6	17.6	"	10.9
4	8.3 8.2	" "	2.3 2.2	9.6 8.2	" "	3.3 1.9	10.8	"	4.5	14.0	"	7.5	9.9	"	3.2	16.5	"	9.8
5	8.0 8.0	" "	2.0 2.0	8.3	"	2.0	8.4 7.3	" "	2.1 1.0	---		---	---		---	15.4	"	8.7
6	8.0 8.5	" —6.1	2.0 2.4	9.5	"	3.2	7.3 9.2	" "	1.0 2.9	---			---		---	10.4	"	3.7
7	9.2	"	3.1	11.8	"	5.5	10.4	"	4.1	---			---			9.0 8.4	" "	2.3 1.7
8	11.7	"	5.6	15.4	"	9.1	12.3	"	6.0	---			---			9.0	"	2.3
9	12.5	"	6.4	15.0	"	8.7	14.2	"	7.9	16.5	—6.6	9.9	12.3	"	5.6	10.4	"	3.7
10	14.4 15.0	" "	8.3 8.9	16.0 16.8	" "	9.7 10.5	15.4	"	9.1	18.4 19.0	" "	11.8 12.4	13.5	"	6.8	12.6	"	5.9
11	15.0 14.2	" "	8.9 8.1	16.8 16.8	" "	10.5 10.5	16.0 16.5	" "	9.7 10.2	19.0 18.4	" "	12.4 11.8	15.2 17.8	" "	8.5 11.1	14.5	"	7.8
Noon	13.4	—6.2	7.2	16.0	—6.3	9.7	17.2 16.9	—6.2 "	11.0 10.7	18.0	—6.7	11.3	18.0 17.4 10.7	" " "	11.3	16.6	"	9.9
1	12.3	"	6.1	14.6	"	8.3	16.4	"	10.0	16.4	"	9.7	16.6	"	9.9	18.9 19.0	" "	12.2 12.3
2	10.3 9.7	" "	4.1 3.5	13.0	"	6.7	13.4	"	7.2	14.4	"	7.7	14.7	"	8.0	19.0 18.1	" "	12.3 11.4
3	8.4 8.4	" "	2.2 2.2	12.1	"	5.8	9.0 8.3	" "	2.8 2.1	12.0	"	5.3	12.5	"	5.8	17.3	"	10.6
4	8.4 9.2	" "	2.2 3.0	9.7 8.6	" "	3.4 2.3	7.2	—6.3	0.9	9.8 8.9	" "	3.1 2.2	10.6	"	3.9	15.5	"	8.8
5	10.3	"	4.1	7.6 9.2	" "	1.3 2.9	7.2 8.1	" "	0.9 1.8	8.0 8.0	" "	1.3 1.3	9.0 8.2	" "	2.3 1.5	12.3	"	5.6
6	11.0	"	4.8	10.3	"	4.0	9.5	"	3.2	9.3	"	2.6	7.5 7.5	" "	0.8 0.8	10.3	"	3.6
7	12.7	"	6.5	12.4	"	6.1	10.7	"	4.4	11.1	"	4.4	8.0	"	1.3	7.9 7.5	" "	1.2 0.8
8	14.5	"	8.3	15.0	"	8.7	13.7	—6.4	7.3	12.9	"	6.2	10.2	"	3.5	7.2 7.2	" "	0.5 0.5
9	15.4	"	9.2	17.4	"	11.1	15.5	"	8.1	14.5	"	7.8	11.8	"	5.1	9.3	"	2.6
10	16.2	"	10.0	19.5 20.0	" "	13.2 13.7	18.2	"	11.8	17.4	"	10.7	14.7	"	8.0	12.1	—6.8	5.3
11	17.1 17.8	" "	10.9 11.6	20.0 19.6	" "	13.7 13.3	19.5	"	13.1	20.3 20.1	" "	13.6 13.4	18.2	"	11.5	16.5	—6.9	9.6
Midn't	18.0	"	11.8	18.8	"	12.5	19.5	"	13.1	20.1	"	13.4	20.2	"	13.5	19.0	—7.0	12.0

May 10 and 11. No sounding.
" 12. Sounding at noon 8 fathoms (last sounding recorded). Correction by sounding and curves.

RECORD AND REDUCTION OF THE TIDES. 31

SERIES III.—TIDAL OBSERVATIONS FROM APRIL 20 TO AUGUST 3, 1854.

Hourly observations on the pulley-gauge. Adopted reading of mean level 7.0, expressed in units of the scale. Increasing numbers indicate rise of water.

May, 1854.

Mean solar hour.	16th.	Ref. obs.	17th.	Ref. obs.	18th.	Ref. obs.	19th.	Ref. obs.	20th.	Ref. obs.	21st.	Ref. obs.	22d.	Ref. obs.	23d.	Ref. obs.	24th.	Ref. obs.
1	19.6 19.6 19.0	12.4 12.4 11.8	18.0	10.8	17.6	10.4	14.0	6.9	12.0 12.3	5.2 5.5	12.0 11.6	5.2 4.8	10.3 10.3 10.3	3.4 3.4 3.4	13.0	5.6	16.2	7.5
2	18.5	11.3	19.4 20.0	11.2 12.8	19.0 19.7	11.8 12.6	15.0	7.9	13.2	6.4	11.2 11.2	4.4 4.4	10.3 10.3 11.3	3.4 3.4 3.8	12.2	4.8	15.1	6.3
3	17.0	9.8	20.0 18.9	12.8 11.7	19.7 19.7	12.5 12.5	16.4	9.3	14.1	7.3	12.5	5.7	10.8 11.5	3.9 4.0	11.0 13.8	3.5 4.9	14.2	5.4
4	16.0	8.8	18.0	10.8	19.7 19.7	12.5 12.5	17.7 18.8	10.6 11.7	15.2	8.5	14.2	7.4	11.5	4.6	12.0 13.0	4.4 4.0	13.0	4.1
5	14.4	7.2	16.2	9.0	19.7 18.6	12.5 11.4	18.8 18.8	11.7 11.7	15.5	8.8	15.6	8.8	13.8	5.9	13.7 13.5	6.1 4.5	13.0	4.0
6	12.5	5.3	14.3	7.1	17.8	10.6	18.8 18.0	11.7 11.0	15.5 16.5	8.8 9.8	16.8 17.4	10.0 10.6	15.2 16.5	8.2 9.5	15.2	7.5	14.2	5.1
7	10.0 9.5	2.8 2.3	12.6	5.4	15.8	8.6	17.4	10.4	16.5 16.5	9.8 9.8	17.9 17.9	11.1 11.1	17.6 17.6	10.6 10.6	17.2 18.2	9.5 10.4	16.6 17.8	7.5 8.6
8	9.0 9.7	1.8 2.5	11.3 10.2	4.1 3.0	14.3	7.1	16.0	9.0	16.5 15.8	9.8 9.1	17.9 17.6	11.1 10.8	17.6 17.6	10.6 10.6	19.1 19.1	11.3 11.2	18.6 18.1	9.4 9.2
9	10.0	2.8	9.0 10.2	1.8 3.0	13.0	5.8	14.4	7.4	14.9	8.2	17.0	10.2	17.0	10.6	18.7	10.8	17.8	8.5
10	10.9	3.7	11.1	3.9	10.0 9.0	2.8 1.8	12.4	5.4	13.6	6.9	15.7	8.9	16.5	9.5	17.5	9.5	16.7	7.4
11	12.4	5.2	13.4	6.2	9.0 10.3	1.8 3.1	11.7 10.6	4.7 3.6	12.8	6.1	14.6	7.8	15.1	8.1	16.0	8.0	15.1	5.8
Noon	14.3	7.1	14.6	7.4	11.6	4.4	10.0 10.5	3.0 3.5	12.1	5.4	12.8	6.0	13.7	6.7	15.3	7.3	14.5	5.2
1	16.2 17.0	9.0 9.8	15.9	8.7	14.0	6.8	10.8	3.8	11.5	4.8	11.2 10.8	4.4 4.0	12.0	5.0	14.2	6.2	13.8	4.5
2	17.6 17.6	10.4 10.4	17.6 17.8	10.4 10.6	15.2	8.0	11.8	4.8	11.0 10.5	4.3 3.8	10.5	3.7	11.0 10.4	4.0 3.4	13.0	5.0	13.1	3.8
3	17.6	10.4	17.8 16.7	10.6 9.5	16.3 16.8	9.1 9.6	12.8	5.8	10.5 10.5	3.8 3.8	10.5 11.0	3.7 4.2	9.5 9.5	2.5 2.5	12.0 11.5	3.9 3.4	12.1 12.1	2.8 2.8
4	16.3	9.1	16.0	8.8	16.5 15.8	8.6 5.6	14.2	7.2	10.8	4.2	11.7	4.9	11.0	4.0	11.5 11.5	3.3 3.3	11.3 11.3	2.0 2.0
5	14.4	7.2	14.0	6.8	15.6	8.4	15.7 16.5	8.7 9.5	12.7	6.0	12.6	5.8	11.7	4.7	12.4	4.0	12.5	3.2
6	12.0 10.8	4.8 3.6	12.8	5.6	14.0	6.8	17.2 17.2	10.2 10.2	14.3	7.6	13.7	6.9	13.1	6.1	14.0	5.6	14.2	4.9
7	9.5 9.5	2.3 2.3	10.6 10.0	3.4 2.8	13.3	6.1	16.5	9.5	14.9 15.2	8.2 8.5	14.5 15.0	7.7 8.2	14.1	7.0	15.5	7.0	16.5	7.3
8	9.8	2.6	9.2 9.2	2.0 2.0	12.4 12.0	5.2 4.9	15.7	8.7	15.4 15.4	8.7 8.7	15.4 15.4	8.6 8.6	10.4 16.5	9.3 9.3	16.6	8.1	18.4	9.2
9	11.7	4.5	10.4	3.2	11.3 11.3	4.2 4.2	15.0	8.1	15.4 14.6	8.7 7.9	15.4 15.2	8.6 8.4	16.5 16.5	9.3 9.3	18.3 19.4	9.7 10.8	19.4	10.2
10	13.5	6.3	12.0	4.8	11.5 11.8	4.4 4.7	14.0	7.1	14.2	7.4	14.7	7.9	16.5 16.5	9.3 9.2	20.0 20.0	11.4 11.4	20.1	10.9
11	15.3	8.1	13.6	6.4	12.4	5.3	13.0	6.1	13.3	6.5	14.0	7.2	16.3	8.0	19.2 22.1	10.6 12.9	21.4 22.1	12.2 12.9
Midn't	16.7	9.5	14.8	7.6	12.8	5.7	12.5	5.7	12.4	5.6	12.6	5.8	15.0	7.7	18.0	9.3	22.1	12.9

From about the middle of May to the end of the series the corrections change very little from day to day, and are given below:—

May 16. Correction —7.2
  " 17.   "   —7.2
  " 18.   "   —7.2
  " 19.   "   —7.0
  " 20.   "   —6.7

May 21. Correction —6.8
  " 22.   "   —7.0
  " 23.   "   —8.0
  " 24.   "   —9.3

32 RECORD AND REDUCTION OF THE TIDES.

SERIES III.—TIDAL OBSERVATIONS FROM APRIL 20 TO AUGUST 3, 1854.

Hourly observations on the pulley-gauge. Adopted reading of mean level 7.0, expressed in units of the scale. Increasing numbers indicate rise of water.

Mean solar hour.	May, 1854.													June, 1854.			
	25th.	Ref. obs.	26th.	Ref. obs.	27th.	Ref. obs.	28th.	Ref. obs.	29th.	Ref. obs.	30th.	Ref. obs.	31st.	Ref. obs.	1st.	Ref. obs.	2d.	Ref. obs.
1	19.8 19.0	10.6 9.8	17.4 16.6	8.5 7.7	21.2 21.2 19.6	12.5 12.5 10.9	19.0 19.0 18.6	10.7 10.7 10.4	18.4	10.8	19.0 18.5 18.6	11.5 12.0 11.1	16.4	8.9	18.2 18.4	10.7 10.9	15.4	8.0
2	17.0	7.9	14.3	5.4	18.7	10.1	17.3	9.1	16.4	8.8	18.1	10.6	17.9 18.5	10.4 11.0	18.5 18.5	11.0 11.0	16.1	8.7
3	14.5	5.4	12.4 11.8	3.6 3.0	16.0	7.4	15.5	7.4	14.3	6.7	17.2	9.5	19.0 19.0	11.5 11.5	18.2	10.7	17.0 17.2	9.6 9.8
4	12.0 11.0	2.9 2.0	11.5 11.5	2.7 2.7	14.7	6.1	12.4	4.3	12.1	4.5	16.1	8.6	18.2	10.7	17.8	10.3	17.2 17.2	9.8 9.8
5	11.0 11.0	2.0 2.0	11.5 11.5	2.7 2.7	12.5 11.2	3.9 2.6	10.8 10.0	2.8 2.0	10.0 9.0	2.5 1.5	14.3	6.8	17.2	9.7	16.2	8.7	17.0	9.7
6	11.5	2.5	11.5 11.8	2.7 3.0	11.0 11.0	2.4 2.4	9.5 9.5	1.6 1.6	9.0 9.0	1.5 1.5	12.4	4.9	14.9	7.4	14.3	6.8	15.6	8.3
7	13.5	4.5	12.5	3.7	12.4	3.8	9.5 9.5	1.6 1.6	9.0 9.0	1.5 1.5	10.2 10.0	2.7 2.5	12.9	5.4	13.4	5.9	14.2	6.9
8	15.1	6.1	14.3	5.5	14.1	5.5	10.1	2.3	9.0 9.0	1.5 1.5	10.0 10.0	2.5 2.5	12.0 11.2	4.5 3.7	12.6 12.0	5.1 4.5	12.8	5.5
9	17.4 18.2	8.4 9.2	15.2	6.4	15.2	6.7	11.7	3.9	10.2	2.7	10.0 11.3	2.5 3.8	10.0 10.0	2.5 2.5	11.5 11.5	4.0 4.0	12.0 11.3	4.7 4.0
10	19.0 19.0	10.0 10.0	16.3	7.5	17.3 18.2	8.8 9.7	13.0	5.2	12.4	4.9	12.4	4.9	10.5	3.0	11.5 11.5	4.0 4.0	11.0 11.0	3.7 3.7
11	18.4	9.4	17.2	8.4	18.2 18.0	9.7 9.5	14.7 15.4	6.9 7.6	14.4	6.9	13.8	6.3	12.6	5.1	12.0	4.5	11.0 11.5	3.7 4.2
Noon	16.0	7.0	18.4 18.5	9.6 9.7	17.5	9.0	16.5 15.7	8.7 7.9	15.5 16.0	8.0 8.5	14.9	7.4	13.7	6.2	13.0	5.5	11.8	4.5
1	15.0	6.0	18.5 17.7	9.7 8.9	16.6	8.1	15.1	7.3	16.0 15.6	8.5 8.1	15.3 16.0	7.8 8.5	14.8 15.4	7.3 7.9	14.5	7.0	12.4	5.1
2	14.0	5.0	17.1	8.3	14.2	5.7	13.5	5.7	15.1	7.6	16.0 16.0	8.5 8.5	16.0 16.0	8.5 8.5	14.6	7.1	12.8	5.5
3	13.2	4.2	16.0	7.2	12.3 9.6	3.8 1.1	11.7 10.8	3.9 3.1	14.2	6.7	16.0 15.2	8.5 7.7	16.0 15.4	8.5 7.9	15.0 15.5	7.5 8.0	13.5	6.2
4	12.0 11.4	3.0 2.4	14.8	6.0	9.0 9.0	0.5 0.5	10.0 10.0	2.3 2.3	12.3	4.8	14.2	6.7	15.0	7.5	15.5 15.5	8.0 8.0	14.4 15.0	7.2 7.8
5	11.0 11.0	2.0 2.0	13.6 13.0	4.8 4.2	9.5	1.0	10.0 10.0	2.3 2.3	11.6	4.1	12.8 11.2	5.3 3.7	13.5	6.0	15.5 15.2	8.0 7.7	15.0 14.3	7.8 7.1
6	11.0 12.1	2.0 3.1	12.2 12.2	3.4 3.4	10.0	1.5	11.3	3.6	16.2 9.5	2.7 2.0	10.0 10.0	2.5 2.5	12.7	5.2	15.0	7.6	14.0	6.8
7	13.3	4.3	12.2 12.6	3.4 3.9	10.4	1.9	12.3	4.6	9.1 9.1	1.6 1.6	10.0 10.0	2.5 2.5	11.7 11.5	4.2 4.0	14.2	6.8	13.4	6.2
8	15.7	6.8	13.2	4.5	12.0	3.6	14.2	6.5	9.1 10.2	1.6 2.7	11.0	3.5	11.0 11.0	3.5 3.5	13.0	5.6	12.8	5.6
9	17.4	8.5	15.3	6.6	14.2	5.8	15.0	7.3	11.3	3.8	13.0	5.5	11.0 11.5	3.5 4.0	12.0 11.4	4.6 4.0	12.4 12.0	5.2 4.8
10	19.5 19.7	10.6 10.8	17.6	8.9	16.4	8.0	16.1	8.4	13.7	6.2	14.4	6.9	12.0	4.5	11.0 11.5	3.6 4.1	12.0 12.0	4.8 4.8
11	19.7	10.8	19.4 20.1	10.7 11.4	18.1 18.7	9.8 10.4	17.4 18.0	9.7 10.3	16.2	8.7	15.1	7.6	14.2	6.7	12.2 12.3	4.8 5.1	12.0	4.8
Mid'n't	19.7	10.8	20.1	11.4	19.0	10.7	18.6	10.9	17.4	9.9	16.2	8.7	15.8	8.3	14.3	6.9	12.7	5.5

May 25. Correction 3.0
 " 26.   "   —8.8
 " 27.   "   —8.5
 " 28.   "   —7.5
 " 29.   "   —7.5

May 30. Correction —7.5
 " 31.   "   —7.5
June 1.   "   —7.5
 "   2.   "   —7.3

RECORD AND REDUCTION OF THE TIDES.

SERIES III.—TIDAL OBSERVATIONS FROM APRIL 20 TO AUGUST 3, 1854.

Hourly observations on the pulley-gauge. Adopted reading of mean level 7.0, expressed in units of the scale. Increasing numbers indicate rise of water.

June, 1854.

Mean solar hour.	3d.	Ref. obs.	4th.	Ref. obs.	5th.	Ref. obs.	6th.	Ref. obs.	7th.	Ref. obs.	8th.	Ref. obs.	9th.	Ref. obs.	10th.	Ref. obs.	11th.	Ref. obs.
1	14.2	7.1	13.0	6.3	12.4	6.0	12.5	6.1	12.3	5.6	12.8	5.5	---	---	16.4	8.2	18.0 19.0	9.5 10.5
2	15.8	8.7	14.0	7.4	12.8	6.4	12.5 12.9	6.1 6.5	12.5	5.8	11.8 11.4	4.4 4.0	---	---	14.9	6.7	19.9 18.0	11.4 9.5
3	16.4	9.3	15.2	8.6	13.8	7.4	13.6	7.2	12.0	5.3	11.4	4.0	---	---	12.1	3.9	15.2	6.7
4	16.6	9.6	16.0	9.4	15.2	8.8	14.2	7.8	11.0	4.3	14.5	7.0	---	---	12.0	3.7	13.2	4.6
5	---	---	---	---	15.4	9.0	14.8	8.4	15.6	8.8	14.7	7.4	---	---	10.4	2.1	10.4	1.8
6	---	---	---	---	15.8	9.4	15.0	8.6	16.0	9.2	15.0	7.7	---	---	10.9	2.6	10.4	1.8
7	---	---	---	---	16.0	9.6	15.6	9.2	17.0	?	15.6	8.2	---	---	11.5	3.2	10.4	1.8
8	---	---	---	---	16.1	9.7	16.0	9.6	15.5	8.7	16.8	9.3	---	---	11.1	?	13.2	4.6
9	11.5	4.5	---	---	14.1	7.7	16.5 15.5	10.1 9.1	17.1 16.8	10.2 9.9	16.0	9.1	---	---	---	---	16.0	7.4
10	12.0	5.0	---	---	12.2	5.8	14.8	8.4	15.5	8.5	16.2	8.6	16.0	8.0	20.0 19.2	11.7 10.9	19.0	9.4
11	12.5	5.5	---	---	11.5 11.3	5.1 4.9	14.5	8.1	14.5	7.5	13.4	5.8	13.9	5.9	18.0	9.7	18.0	9.4
Noon	13.0	6.0	---	---	11.1 11.1	4.7 4.7	13.3	6.9	12.9	5.9	12.5	4.9	12.6	4.6	17.6	9.3	17.2	8.6
1	13.5	6.5	---	---	11.1 11.2	4.7 4.8	11.6 10.8	5.2 4.4	11.9	4.9	11.8 10.4	4.2 2.7	12.0 10.3	4.0 2.3	16.4	8.1	16.8	8.2
2	14.0	7.0	11.4	5.0	11.2	4.8	10.5 10.5	4.0 4.0	11.4 10.5	4.4 3.5	9.5	1.8	11.0	3.0	14.7	6.4	16.0	7.4
3	14.5 15.0	7.6 8.1	12.4	6.0	11.4	5.0	10.5 10.9	4.0 4.4	10.5 11.5	3.4 4.4	9.5 9.8	1.8 2.1	9.4	1.4	13.2	4.9	15.0	6.4
4	15.0 15.0	8.1 8.1	13.6	7.2	12.4	6.0	11.4	4.9	12.6	5.5	10.5	2.8	10.3	2.3	11.7	3.3	14.2	5.6
5	14.8	6.0	14.4	8.0	13.6	7.2	13.2	6.6	14.0	6.9	11.9	4.1	11.4	3.4	10.2 9.6	1.8 1.2	12.9	4.3
6	14.6	7.8	14.4	8.0	14.8	8.4	14.8	8.2	14.7	7.5	13.1	5.3	12.6	4.5	9.0	0.6	12.0 10.6	3.4 2.0
7	14.0	7.2	15.0	8.6	14.4	8.0	16.2	9.6	17.0	9.8	14.0	6.2	16.4	8.3	12.0	3.6	11.5	2.8
8	13.4	6.6	15.0	8.6	15.2	8.8	17.0	10.4	18.0	10.8	16.2	8.4	19.5	11.4	15.4	7.0	13.6	4.9
9	13.0 12.0	6.2 5.2	14.0	7.6	15.6	9.2	16.6	10.0	18.4	11.1	18.6	10.8	21.0	12.9	---	---	16.2	7.5
10	12.0 12.0	5.2 5.2	13.0	6.6	14.6	8.2	16.2	9.6	18.0	10.7	19.4	11.6	20.6	12.5	---	---	19.2	10.5
11	12.3	5.6	12.5	6.1	13.2	6.8	16.2	9.6	17.2	9.9	19.0	11.2	20.0	11.9	---	---	20.2	11.5
Midn't	13.0	6.3	12.0	5.6	12.6	6.2	15.2	8.6	14.0	6.7	18.0	10.2	---	---	---	---	21.4	12.6

June 3. Correction —7.0   June 4. Correction —6.4   June 5. Correction —6.4
" 6. " —6.4   " 7. " —7.0   " 8. " —7.6
" 9. " —8.0   The record on this day is defective.
" 10. " —8.3   Readings between the full hours are less frequent than before, and are generally
" 11. " —8.6   [given only near high or low water.

SERIES III.—TIDAL OBSERVATIONS FROM APRIL 20 TO AUGUST 3, 1854.

Hourly observations on the pulley-gauge. Adopted reading of mean level 7.0, expressed in units of the scale. Increasing numbers indicate rise of water.

June, 1854.

Mean solar hour.	12th.	Ref. obs.	13th.	Ref. obs.	14th.	Ref. obs.	15th.	Ref. obs.	16th.	Ref. obs.	17th.	Ref. obs.	18th.	Ref. obs.	19th.	Ref. obs.	20th.	Ref. obs.
1	22.0	13.2	21.0 21.0	12.1 12.1	21.0 21.0	12.1 12.1	19.2 20.1	10.4 11.3	16.0	7.3	17.3	8.6	14.4	6.1	14.0	6.2	13.5 13.0	5.8 5.3
2	21.5 19.8	12.7 11.0	19.9	11.0	21.0	12.1	21.1 21.1	12.3 12.3	17.2	8.5	18.6 19.2	9.9 10.5	15.6	7.3	15.1	7.4	12.9 12.9	5.2 5.2
3	18.5	9.7	18.2	9.3	20.2	11.3	21.1 20.9	12.3 12.1	19.0 20.0	10.3 11.3	20.0	11.4	17.0	8.7	15.9	8.2	14.0	5.3
4	15.2	6.4	18.0	9.1	19.6	10.7	20.2	11.5	20.4	11.7	19.0	11.3	18.2	9.9	16.2	8.6	14.0	6.2
5	12.5	3.7	13.2	4.3	15.3	6.4	16.3	7.6	19.0	10.3	17.6	9.0	18.4	10.2	18.0	10.4	15.6	7.8
6	10.4	1.6	11.6	2.7	12.2	3.3	14.4	5.7	16.8	8.1	16.4	7.8	18.0	9.8	18.1	10.5	16.4	8.6
7	10.4	1.6	10.0	1.1	10.5	1.6	11.5	2.8	14.3	5.6	14.2	5.6	17.5	9.3	18.0	10.4	17.6	9.8
8	13.2	4.4	10.0 9.1	1.1 0.2	10.5	1.6	10.8 10.4	2.1 1.7	12.8	4.1	12.4 11.6	3.8 3.0	17.0	8.8	17.8	10.2	17.7	9.9
9	15.5	6.7	10.0	1.1	12.6	3.7	10.0 11.0	1.3 2.3	11.4 11.2	2.7 2.5	11.4	2.8	13.4	5.3	15.6	8.0	17.6	9.8
10	16.6	7.8	12.2	3.3	14.0	5.1	12.0	3.3	11.4	2.5	12.2	3.6	12.5	4.4	15.0	7.1	16.5	8.7
11	18.6 19.2	9.8 10.4	16.4	7.5	16.0	7.1	14.5	5.8	13.0	4.3	12.6	4.0	11.6 11.2	3.6 3.2	14.0	6.4	14.8	7.0
Noon	17.6	8.8	18.4 19.2 19.2	9.5 10.3 10.3	17.6 19.2	8.7 10.3	16.2 16.5	7.5 7.8	14.6	5.9	14.5	5.9	11.4	3.4	13.0	5.4	14.5	6.7
1	15.8	7.0	19.2 19.2	10.3 10.3	19.1	10.2	17.2 17.2	8.5 8.5	15.9 16.8	7.2 8.1	15.9	7.3	12.0	4.0	12.0 11.6	4.4 4.0	14.0	6.2
2	13.8	5.0	18.0	10.0	19.1	10.2	17.2 16.2	8.5 7.5	17.6 16.9	8.9 8.2	17.3 17.9	8.7 9.3	13.3	5.3	12.0	4.4	13.6	5.8
3	13.0	4.2	17.6	8.7	17.4	8.5	15.5	6.8	16.2	7.5	17.5	8.9	16.6	8.6	13.0	5.4	13.0 12.2	5.2 4.4
4	11.6	2.5	16.2	7.3	16.6	7.1	15.0	6.3	15.5	6.8	17.0	8.4	16.3 16.0	8.3 8.0	14.0	6.4	12.4	4.6
5	10.3 9.6	1.5 0.8	14.8	5.9	14.2	5.3	13.4	4.7	14.8	6.1	16.6	8.0	18.5 15.0	10.5 7.0	15.0	7.4	14.0	6.1
6	9.2 9.6	0.4 0.8	14.2	5.3	13.0 10.4	4.1 1.5	12.6	3.9	14.6	5.9	16.1	7.5	14.9	6.3	16.5	8.9	15.0	7.1
7	10.2	1.4	14.0	5.1	10.4 10.4	1.5 1.5	11.6 11.2	3.9 2.5	14.2	5.5	15.8	7.2	13.2	5.2	17.4 17.6	8.8 10.0	16.7	8.8
8	12.4	3.6	10.0 11.0	1.1 2.1	10.2 10.6	1.3 1.7	11.0 11.2	2.3 2.5	12.8	4.1	15.2	6.7	13.6	5.6	17.4	9.8	17.8	9.9
9	15.0	6.2	12.0	3.1	11.0	2.1	11.4	2.7	12.4	3.7	14.2	5.7	14.3	6.3	16.8	9.1	18.4	10.5
10	18.5	9.7	15.0	6.1	12.4	3.6	13.6	4.9	13.0	4.3	13.4	4.9	15.0	?	15.5	7.8	18.5	10.5
11	21.0	12.3	18.6	9.1	15.0	6.2	15.0	6.3	14.0	5.3	13.0	4.6	14.0	6.2	14.5	6.8	16.9	8.9
Midn't	---	---	20.4	11.5	18.0	9.2	---	---	15.2	6.5	13.0	4.6	13.4	5.6	14.3	6.6	15.6	7.6

June 12. Correction —8.5 The record on this day is defective, the times being uncertain.
" 13.   "   —8.9   June 14. Correction —8.9
" 15.   "   —8.7   " 16.   "   —8.7
" 17.   "   —8.6   " 18.   "   —8.0
" 19.   "   —7.6   " 20.   "   —7.8

RECORD AND REDUCTION OF THE TIDES. 35

SERIES III.—TIDAL OBSERVATIONS FROM APRIL 20 TO AUGUST 3, 1854.

Hourly observations on the pulley-gauge. Adopted reading of mean level 7.0, expressed in units of the scale. Increasing numbers indicate rise of water.

June, 1854.

Mean solar hour.	21st.	Ref. obs.	22d.	Ref. obs.	23d.	Ref. obs.	24th.	Ref. obs.	25th.	Ref. obs.	26th.	Ref. obs.	27th.	Ref. obs.	28th.	Ref. obs.	29th.	Ref. obs.
1	14.5	6.5	15.2	6.6	16.5	7.0	21.0 20.5	10.6 10.1	19.0	8.3	22.4 22.4	12.1 12.1	20.2	10.1	21.4	11.4	21.4 21.4 21.4	11.4 11.4 11.4
2	14.2 13.8	6.2 5.8	13.9	5.3	14.7	5.2	18.0	7.6	17.0	6.3	20.0	9.7	20.2	10.1	21.4	11.4	21.4	11.4
3	13.8 13.8 13.8	5.8 5.8 5.8	11.5 13.3	2.9 4.7	13.8 13.4	4.3 3.8	15.8 15.0	5.3 4.5	16.0	5.3	17.6	7.3	18.5	8.4	19.1	9.1	20.0	10.0
4	14.0	5.9	---	---	13.2	3.6	14.2	3.6	14.1	3.5	15.0	4.8	17.0	6.9	16.3	6.3	18.0	8.0
5	14.4	6.3	---	---	14.0	4.3	14.2	3.6	13.0	2.4	12.2	2.0	12.0	1.9	15.3	5.3	16.6	6.6
6	15.7	7.6	---	---	14.6	4.9	15.1	4.4	13.4	2.8	13.4	3.2	12.2	2.1	14.4	4.4	15.5	5.5
7	16.3	8.2	---	---	15.0	5.2	15.3	4.6	14.0	3.4	13.8	3.7	12.8	2.7	13.8	3.8	13.9	3.9
8	17.2 17.2	9.1 9.0	16.7	7.6	15.6	5.8	15.5	4.7	14.5	3.9	14.4	4.3	13.4	3.3	13.2 13.5	3.2 3.5	13.6 13.9	3.6 3.9
9	17.6 17.2	9.4 9.0	17.6 19.0	8.5 9.9	18.4	8.5	16.4	5.6	16.2	5.6	16.0	5.9	14.6	4.5	14.0	4.0	14.4	4.4
10	16.2	8.0	18.5	9.4	18.4	8.5	17.5	6.7	17.0	6.4	17.2	7.1	16.4	6.3	15.5	5.5	16.2	6.2
11	15.8	7.6	17.5	8.4	18.1	8.1	18.3 18.5	7.5 7.7	18.1 18.8	7.5 8.2	18.8	8.7	18.0	7.9	17.8	7.8	18.0	8.0
Noon	14.6	6.4	16.5	7.4	17.8	7.8	18.9 18.9	8.1 8.1	18.9	8.3	19.8	9.7	19.4 19.9	9.3 9.8	19.1 19.2	9.1 9.2	18.4	8.4
1	14.0	5.8	16.0	6.9	16.0	6.0	18.5	7.7	18.2	7.6	20.0 19.0	8.9 8.9	19.0	8.9	19.2 19.2	9.2 9.2	19.0 19.1	9.0 9.1
2	13.0 12.4	4.8 4.2	15.4	6.3	15.5	5.5	15.9	5.1	16.5	5.9	18.2	8.1	17.6	7.5	18.6	5.6	19.1 18.8	9.1 8.8
3	11.2 11.0	3.0 3.7	14.4	5.3	14.0	3.9	14.7	3.9	15.7	5.1	16.4	6.3	16.8	6.7	17.5	7.5	17.5	7.5
4	13.9	5.6	13.2 12.8	4.1 3.7	13.6 13.5	3.5 3.3	13.9	3.1	---	---	15.0	4.9	16.0	5.9	17.2	7.2	16.4	6.4
5	14.6	6.3	12.5 13.0	3.3 3.8	13.2 13.2	3.0 2.9	13.2 12.8	2.4 2.0	---	---	14.0	3.9	14.8	4.7	16.8	6.8	13.4	3.4
6	16.5	8.2	14.6	5.4	13.2 14.2	2.9 3.9	13.5	2.7	13.7	3.2	13.3 12.5	3.2 2.4	13.8	3.8	15.1	5.1	12.1 12.0	2.1 2.0
7	18.4	10.0	17.2	7.9	15.0	4.7	15.0	4.2	13.6	3.1	13.0	2.9	13.8 13.0	3.8 3.0	12.8 12.5	2.8 2.5	12.6	2.6
8	19.5	11.1	18.0	8.7	16.0	5.7	17.0	6.2	15.0	4.5	15.0	4.9	13.2	3.2	13.1	3.1	13.5	3.6
9	20.0 21.5	11.6 13.0	19.0 19.5	9.6 10.1	19.0	8.7	19.0	8.2	18.0	7.5	19.4	9.3	15.0	5.0	15.1	5.1	15.2	5.3
10	20.5	12.0	20.0 20.0	10.6 10.6	21.0	10.6	21.0 21.5	10.2 10.8	20.2	9.8	22.2	12.1	17.5	7.5	17.2	7.2	16.7	6.8
11	18.5	10.0	19.8	10.3	21.4 21.4	11.0 11.0	22.0 21.8	11.3 11.1	21.5	11.1	22.4	12.3	20.0	10.0	19.2	9.2	17.8	8.0
Midn't	14.1	5.6	19.4	9.9	21.4	11.0	20.6	9.9	22.0	11.6	---	---	21.4	11.4	21.3	11.3	19.5	9.7

June 21. Correction — 8.2   June 22. Correction — 9.1
" 23.   "   —10.0   " 24.   "   —10.8
" 25.   "   —10.6   " 26.   "   —10.1
" 27.   "   —10.1   Some doubt about the time record in the afternoon.
" 28.   "   —10.0
" 29.   "   —10.0

RECORD AND REDUCTION OF THE TIDES.

SERIES III.—TIDAL OBSERVATIONS FROM APRIL 20 TO AUGUST 3, 1854.

Hourly observations on the pulley-gauge. Adopted reading of mean level 7.0, expressed in units of the scale. Increasing numbers indicate rise of water.

Mean solar hour.	June, 1854.				July, 1854.														
	30th.	Ref. obs.	1st.	Ref. obs.	2d.	Ref. obs.	3d.	Ref. obs.	4th.	Ref. obs.	5th.	Ref. obs.	6th.	Ref. obs.	7th.	Ref. obs.	9th.	Ref. obs.
1	19.0	9.2	18.5	9.0	20.6	11.0	14.5	5.9	16.3	7.6	15.2	6.5	15.7 15.0 15.0	6.7 6.0 6.0	17.0 17.0	7.8 6.8	15.5	8.3
2	20.8	11.0	19.8	10.3	21.0	11.3	16.3	7.7	16.9	8.2	15.6	6.9	15.0 15.0	6.0 6.0	15.8 15.8	5.6 5.6	13.0	5.8
3	21.0 22.0	11.2 12.3	20.7 20.9	11.2 11.4	21.4	11.7	17.3 17.6	6.7 8.9	17.4	8.7	16.8	8.1	15.0 15.6	6.0 6.6	15.8 15.8	5.6 5.6	11.0	3.9
4	20.4	10.7	21.0 21.0	11.5 11.6	21.4	11.7	17.8 17.8	9.1 9.1	18.6	9.9	17.3	8.6	15.6	6.6	15.8 16.0	5.6 5.8	9.8 9.5	2.7 2.4
5	19.8	10.1	21.0 20.7	11.6 11.3	17.6	7.8	17.5	8.8	18.6	10.0	18.4	9.7	17.0	8.0	16.4	6.2	9.5 9.5	2.4 2.5
6	18.0	8.4	19.5	10.1	15.7	5.9	17.0	8.2	18.6 18.8	10.0 10.2	18.6 18.8	9.9 10.0	17.6	8.5	17.2	7.0	9.8	2.8
7	17.0	7.4	17.1	7.7	14.4	4.6	16.3	7.5	18.8 18.4	10.2 9.8	18.9 18.4	10.1 9.6	18.2 18.5	9.1 9.4	18.8 19.0	8.6 8.8	10.5	3.5
8	15.0 14.0	5.4 4.4	15.7	6.3	12.7	4.5	15.7	6.9	17.6	9.0	18.4	9.6	18.5 17.8	9.4 8.7	19.4 19.1	9.2 6.9	12.4	5.4
9	13.5 14.8	3.9 5.2	13.5 12.2	4.1 2.8	12.0 11.7	3.8 3.5	14.0 14.0	5.2 5.2	15.0	6.4	18.2	9.4	17.0	7.8	18.5	8.3	13.8	6.6
10	16.1	6.5	12.0 12.5	2.6 3.1	11.4 11.4	2.2 2.2	13.6 13.6	4.8 4.8	14.0 13.6	5.4 5.0	17.0	8.2	16.2	7.0	16.8	6.6	15.4 15.5	8.4 8.5
11	18.7 18.9	9.1 9.3	13.0	3.6	11.4 11.4	2.2 2.2	13.6 15.0	4.8 6.2	13.0 13.0	4.4 4.4	15.0	6.2	15.9	6.7	15.5	5.3	15.5 14.9	8.5 7.9
Noon	19.0 19.0	9.4 9.4	15.4	6.0	11.9	3.7	16.5	7.7	13.1	4.5	14.0 13.3 13.1	5.2 4.5 4.3	14.0 13.1 13.1	4.8 3.9 3.9	15.5	5.3	14.3	7.3
1	18.8	9.2	17.2	7.8	12.2	3.9	18.3	?	13.2	4.6	13.1 13.1	4.3 4.3	13.1 13.1	3.9 3.9	14.4	4.2	13.2	6.2
2	17.0	7.4	18.1	8.7	13.4	5.1	16.2	7.4	14.0	5.4	13.1 13.2	4.3 4.4	13.1 13.1	3.9 3.9	13.3 13.2	3.1 3.0	11.2	4.3
3	15.8	6.2	18.8	8.9	14.4	6.1	17.2	8.4	15.3	6.7	13.4	4.6	13.4	4.2	13.2 13.2	3.0 3.0	9.1 8.3	2.2 1.5
4	15.0	5.4	18.3	8.9	15.2	6.8	18.2 18.4	9.4 9.6	17.2	8.6	16.1	7.3	14.3	5.1	13.2 13.7	3.0 3.5	8.0 8.0	1.2 1.3
5	14.0	4.4	18.7 17.8	9.3 8.4	15.6 15.7	7.2 7.3	18.6 18.6	9.8 9.8	17.4	8.7	17.5	8.7	17.3	8.1	14.2 15.3	4.0	8.0 8.6	1.3 2.0
6	13.5	3.9	17.3	7.8	15.3	6.9	18.6 18.3	9.8 9.5	18.0 18.5	9.3 9.8	18.0 18.7	9.2 9.9	18.3	9.1	16.6	5.1	9.0	2.4
7	13.4	3.8	16.2	6.7	14.6	6.2	18.0	9.2	18.5 18.5	9.8 9.8	19.4 19.4	10.6 10.5	19.5 19.9	10.3 10.7	17.4	6.4	11.0	4.5
8	13.4	3.8	14.7	5.2	14.0	5.5	17.5	8.7	18.0	9.3	19.4 19.4	10.5 10.5	20.4 20.4	11.2 11.2	18.0	7.2	13.4	6.9
9	14.1	4.5	14.5	5.0	14.0	5.5	16.7	7.9	17.4	8.7	19.4 19.2	10.5 10.3	20.4 20.4	11.2 11.2		7.8	15.9	9.5
10	15.4	5.8	17.5	8.0	12.8 12.4	4.3 3.9	15.8 15.6	7.0 6.9	16.2	7.5	18.7	9.8	20.4 20.1	11.2 10.9	Readings irregular.		16.2	11.9
11	16.0	6.4	18.9	9.3	12.3 12.3	3.8 3.8	15.6 15.6	6.9 6.9	15.8	7.1	17.2	8.3	19.7	10.5			18.6 18.7	12.3 12.5
Midn't	16.7	7.1	19.7	10.1	12.3	3.8	15.7	7.0	15.2	6.5	16.0	7.1	17.1	7.9			18.7	12.5

June 30. Correction —9.6   July 1. Correction —9.4
July 2.   "   —9.8 before 8 A. M., and —8.2 after 8 A. M.   " 3.   "   —8.8
 "   4.   "   —8.6   " 5.   "   —8.8
 "   6.   "   —9.2
 "   7.   "   —9.2 Tide register out of order at 2 o'clock, changed index 1 foot; correction after
 "   8. The readings appear irregular. Correction at noon —7.0, at midnight —6.2.   [2 A. M. 10.2.

RECORD AND REDUCTION OF THE TIDES. 37

SERIES III.—TIDAL OBSERVATIONS FROM APRIL 20 TO AUGUST 3, 1854.

Hourly observations on the pulley-gauge. Adopted reading of mean level 7.0, expressed in units of the scale. Increasing numbers indicate rise of water.

July, 1854.

Mean solar hour.	10th.	Ref. obs.	11th.	Ref. obs.	12th.	Ref. obs.	13th.	Ref. obs.	14th.	Ref. obs.	15th.	Ref. obs.	18th.	Ref. obs.	19th.	Ref. obs.	20th.	Ref. obs.
1	18.4 16.8	12.3 10.7	19.3 19.0	13.4 13.1	19.3 19.3 19.3	13.3 13.3 13.3	19.5 19.6 19.6	13.5 13.6 13.6	17.0 18.0	11.0 12.0	18.1 16.1	? "	---	---	---	---	20.8	6.1
2	14.1	8.1	17.0	11.0	19.0	13.0	19.0	13.0	18.6	12.6	13.2	"	---	---	---	---	20.5 20.4	5.8 5.6
3	12.5	6.5	14.0	8.0	16.5	10.5	17.8	11.8	18.4	12.4			---	---	---	---	20.4 20.5	5.6 5.7
4	10.0 9.0	4.0 3.0	12.1	6.1	12.0	6.0	16.3	10.3	17.3	11.3			---	---	---	---	20.8	5.9
5	8.2 8.2	2.2 2.2	9.1	3.1	10.1	4.1	14.0	8.0	15.0	9.0		?			19.1 19.2	4.6 4.7	20.4	5.5
6	8.2 8.5	2.2 2.5	8.0 7.5	2.0 1.5	7.8 7.1	1.8 1.1	10.2	4.2	---	---			irregular.	"	19.4	4.8	21.2	6.3
7	9.0 9.4	3.0 .?	7.8 8.5	1.8 2.5	7.0 7.3	1.0 1.3	7.5 7.0	1.5 1.0	---	---			Readings	"	20.3	5.7	22.1	7.2
8	6.6	---	9.0	3.0	8.2	2.2	7.0 7.1	1.0 1.1	---	---				"	21.2	6.6	22.2	7.3
9	7.2	---	11.5	6.5	10.5	4.5	8.0	2.0	7.8 7.8	1.8 1.8				"	22.2 21.7	7.6 7.1	23.0 23.3	8.1 8.4
10	10.3	4.5	14.0	8.0	13.0	7.0	10.2	4.2	7.9	1.9				"	21.2	6.6	23.6 23.5	8.7 8.6
11	12.1	6.3	15.6	9.6	15.4 16.1	9.4 10.1	12.6	6.6	9.3	3.3			irregular.	"	21.8	7.2	23.5	8.6
Noon	13.5 14.0	7.7 8.2	16.0 16.1	10.0 10.1	16.1 16.0	10.1 10.0	14.9 15.9	8.9 9.9	12.3	6.3			very	21.3 20.2	7.0 5.9	20.0	5.4	22.2	7.3
1	14.2 13.1	8.4 7.3	16.0	10.0	16.0	10.0	16.0 16.0	10.0 10.0	14.1 14.4	8.1 8.4			become	19.1 19.1	4.8 4.8	19.6 19.0	5.0 4.4	20.1	5.2
2	12.1	6.3	15.2	9.2	15.3	9.3	16.0 15.1	10.0 9.1	15.2 15.2	9.2 9.2			Readings	19.1 19.3	4.8 5.0	19.3	4.7	19.4 19.7	4.5 4.8
3	10.2	4.4	13.2	7.2	13.2	7.2	14.2	8.2	15.2 14.4	9.2 8.4				20.2	5.9	19.3	4.7	20.4	5.5
4	8.0 7.8	2.2 2.0	10.0	4.0	11.0	5.0	13.3	7.3	14.3	8.3				21.3	7.0	20.0	5.4	21.2	6.3
5	7.0 7.0	1.2 1.2	8.7 7.8	2.7 1.8	10.0	4.0	---	---	13.0	7.0				22.1	7.7	21.1	6.5		
6	7.6	1.7	7.2 7.2	1.2 1.2	8.7 8.0	2.7 2.0	---	---	12.8	6.8				23.2 23.4	8.8 9.0	22.4	7.8		
7	10.1	4.2	7.5	1.5	7.2 7.2	1.2 1.2	9.0 8.7	3.0 2.7	---	---				23.6 23.6	9.2 9.2	23.6	9.0		irregular.
8	13.1	7.2	9.0	3.0	7.6	1.6	8.4	2.4	---	---				23.6 23.3	9.2 8.9	23.0 24.0	9.2 9.3		
9	15.0	9.1	10.5	4.5	9.2	3.2	---	---	8.5 8.9	2.5 2.9				22.9	8.5	24.0 23.4	9.3 8.7		Readings
10	17.0	11.1	13.9	7.9	12.2	6.2	---	---	9.0	3.0				22.1	7.7	23.4	8.7		
11	18.6 19.0	12.7 13.1	16.8	10.8	16.8	10.8	---	---	11.5	5.5				21.4	7.0	23.6	8.7		
Midn't	19.5	13.6	19.0	13.0	18.2	12.2	---	---	13.2	7.2				20.3	5.9	21.3	6.6		

```
July 10. Correction — 5.8          July 11. Correction — 6.0
  "  12.     "      — 6.0            "   13.     "      — 6.0
  "  14.     "      — 6.0            "   15.     "         ?
  "  18.     "      —14.3            "   19.     "      —14.6
  "  20.     "      —14.9
```

SERIES III.—TIDAL OBSERVATIONS FROM APRIL 20 TO AUGUST 3, 1854.

Hourly observations on the pulley-gauge. Adopted reading of mean level 7.0, expressed in units of the scale. Increasing numbers indicate rise of water.

	July, 1854.								August, 1854.					
Mean solar hour.	28th.	Ref. obs.	29th.	Ref. obs.	30th.	Ref. obs.	31st.	Ref. obs.	1st.	Ref. obs.	2d.	Ref. obs.	3d.	Ref. obs.
1			10.6	10.1	10.6	10.5	8.3	8.3	8.0	8.0	6.8	7.0	5.3	5.7
					11.2	11.1							5.5	5.9
2			11.3	10.8	11.2	11.1	8.6	8.6	8.4	8.4	8.0	8.2	6.0	6.4
			12.3	11.8	11.2	11.1			9.1	9.1				
3	Correction to readings not sufficiently known.		12.5	12.0	11.2	11.1	9.3	9.3	9.3	9.3	8.7	8.9	6.5	6.9
			11.3	10.8	11.0	11.0								
4			10.2	9.7	10.8	10.8	10.6	10.6	9.3	9.3	9.0	9.3	7.3	7.7
									9.2	9.2			8.2	8.6
5			8.4	8.0	10.2	10.2	---	---	9.0	9.0	10.2	10.5	8.4	8.8
									9.0	9.0			8.4	8.8
6			6.1	5.7	8.5	8.5	---	---	9.0	9.0	9.1	9.4	8.4	8.8
													8.4	8.8
7			3.0	2.6	6.4	6.4	---	---	7.5	7.5	7.4	7.8	8.4	8.8
			2.3	1.9	5.1	5.1								
8			2.0	1.7	3.2	3.2	---	---	5.5	5.5	6.4	6.8	7.7	8.1
			2.0	1.7	3.2	3.2								
9			2.2	1.9	3.2	3.2	4.0	4.0	4.6	4.6	5.0	5.4	6.0	6.4
					3.3	3.3			4.4	4.4				
10			3.5	3.2	3.4	3.4	5.2	5.2	4.4	4.4	4.2	4.6	4.4	4.8
									4.4	4.4	4.1	4.5		
11			4.8	4.5	4.2	4.2	5.5	5.5	4.4	4.4	4.1	4.5	3.4	3.8
									4.4	4.4	4.1	4.5	3.4	3.8
Noon	8.3	7.6	7.2	6.9	6.1	6.1	6.2	6.2	4.4	4.4	4.1	4.5	3.4	3.8
	9.1	8.4							5.0	5.0	4.3	4.7	3.2	3.6
1	9.2	8.5	9.2	8.9	7.0	7.0	7.5	7.5	5.7	5.7	4.8	5.2	3.2	3.6
	9.3	8.6					8.0	8.0					3.2	3.6
2	9.3	8.6	10.2	9.9	8.3	8.3	9.0	9.0	7.0	7.0	5.9	6.3	3.2	3.6
	9.3	8.6			8.5	8.5	9.0	9.0					4.0	4.4
3	9.2	8.5	11.3	10.0	9.2	9.2	9.0	9.0	8.7	8.7	6.4	6.8	4.5	4.9
			11.4	11.2	9.0	9.0	9.0	9.0	9.5	9.5				
4	7.4	6.8	11.5	11.3	8.3	8.3	9.0	9.0	10.4	10.5	6.8	7.2	6.2	6.6
			11.0	10.8					10.4	10.5				
5	5.5	4.9	10.0	9.8	7.2	7.2	9.1	9.1	10.4	10.5	7.0	7.4	7.4	7.8
									10.4	10.5	8.2	8.6		
6	3.7	3.1	7.8	7.6	6.0	6.0	8.0	8.0	10.2	10.3	9.1	9.5	8.2	8.6
											9.0	9.4		
7	2.9	2.3	4.2	4.1	5.1	5.1	6.6	6.6	9.0	9.1	9.0	9.4	8.8	9.2
	2.7	2.1			4.3	4.3							9.0	9.4
8	2.6	2.0	3.4	3.3	4.1	4.1	5.4	5.4	7.4	7.5	9.0	9.4	9.0	9.4
	3.0	2.4			4.1	4.1	4.8	4.8					9.0	9.4
9	3.2	2.7	3.3	3.2	4.1	4.1	4.8	4.8	6.2	6.3	7.4	7.8	9.0	9.4
			4.0	3.9	5.0	5.0	4.8	4.8	6.0	6.1			9.0	9.4
10	5.4	4.9	4.0	3.9	5.2	5.2	4.8	4.8	6.0	6.1	6.2	6.6	8.2	8.6
							4.8	4.8	6.0	6.1				
11	6.4	5.9	6.0	5.9	6.4	6.4	4.8	4.8	6.2	6.4	5.8	6.2	7.3	7.7
							5.1	5.1						
Midn't	9.0	8.5	8.0	7.9	---	---	5.6	5.6	6.5	6.7	5.6	6.0	6.0	6.4

Between the 20th and 27th of July the observations do not appear sufficiently regular to promise any reliable results.

July 28. Correction —0.7   July 29. Correction —0.3   July 30. Correction 0.0
" 31. " 0.0   Aug. 1. " —0.0   Aug. 2. " +0.4
Aug. 3. " +0.4   After this date the observations are irregular.

On the 5th the rope slipped off the wheel.

Aug. 8. The brig was released from the ice cradle at 10 A. M., rising suddenly 2½ feet. She resumed this position upon very slight disturbance of the external ice, and is now on an even keel for the first time in eleven months. The brig was frozen in and fast since the 9th of September, 1853.

Aug. 10. The high-water mark was cut on the island by Mr. McGary.

Aug. 11. The warping of the ships was commenced. Tidal observations were resumed on the 12th. The register is kept in fathoms and feet.

Series IV.—Tidal Observations from September 7 to October 22, 1854.

Hourly observations on the pulley-gauge. Adopted reading of mean level 7.0, expressed in units of the scale. Increasing numbers indicate rise of water.

September, 1854.

Mean solar hour.	7th.	8th.	9th.	10th.	11th.	12th.	13th.	14th.	16th.	17th.	18th.	19th.	20th.	21st.	22d.	23d.	24th.	25th.	26th.	27th.
1	---	10.0	13.5	13.0	10.0	10.0	8.0	6.0	---	5.0	7.0	9.0	9.0	10.0	11.2	12.0	11.0	13.0	13.0	10.0
2	---	5.0	13.5	11.0	11.0	14.0	9.0	7.0	---	5.0	6.0	8.0	7.5	8.0	9.0	10.0	10.0	14.0	11.0	12.0
3	---	2.0	11.0	7.0	11.0	11.0	10.0	4.0	---	6.0	6.0	7.0	5.0	6.0	4.0	7.0	8.0	10.0	10.0	10.0
4	---	—1.0	7.0	4.7	9.0	9.0	9.0	3.0	---	7.0	5.0	5.0	3.0	5.0	3.0	3.0	6.0	7.0	8.0	8.0
5	---	—1.7	5.0	3.0	7.0	6.0	6.0	4.0	---	8.0	6.0	5.0	2.0	4.0	2.0	0.0	3.5	4.0	6.0	4.0
6	---	—1.7	3.0	1.0	5.0	4.0	4.0	5.0	---	9.0	7.0	7.0	4.0	5.0	3.0	—0.7	2.0	3.0	3.0	2.0
7	---	—1.0	2.2	—1.0	3.0	2.0	0.0	6.0	---	9.0	9.0	8.0	6.0	6.0	5.0	2.0	2.0	2.0	0.0	0.0
8	---	0.0	1.0	—1.0	0.0	0.0	1.0	8.0	---	9.0	8.5	10.0	7.0	7.0	6.0	4.0	2.0	0.6	1.0	—1.0
9	---	2.0	4.0	0.0	0.0	1.0	2.0	9.0	---	8.0	7.5	10.0	8.0	8.5	8.0	7.0	5.0	1.0	3.0	1.0
10	---	9.0	5.0	4.0	2.0	3.0	3.0	10.0	---	6.0	9.0	9.5	9.0	10.0	10.0	10.0	10.0	3.0	4.0	3.0
11	---	14.0	6.5	---	4.0	4.0	6.0	4.0	4.0	5.0	7.0	8.5	9.5	11.5	12.5	11.0	12.0	6.0	7.0	6.0
Noon	---	12.0	11.0	---	7.0	7.0	6.0	---	4.0	5.0	6.5	8.0	---	10.0	13.0	13.0	13.0	12.5	10.0	9.0
1	---	11.0	13.0	9.0	9.0	3.0	8.0	---	5.0	5.0	6.0	7.5	---	9.0	10.0	11.0	12.0	13.0	12.0	10.0
			14.0																	
2	---	10.0	10.0	11.0	10.0	4.0	9.0		7.0	6.0	5.5	7.0	---	6.0	8.0	10.0	10.0	12.0	13.0	12.0
3	---	8.0	7.0	9.0	8.0	9.0	9.0		8.0	7.0	5.0	6.7	4.0	0.0	6.0	7.0	9.0	11.0	11.0	13.0
4	---	6.0	4.0	6.0	9.0	10.0	9.0		9.0	8.0	7.0	6.5	2.0	1.0	3.0	5.0	6.0	7.0	9.0	10.0
5	---	4.0	2.0	---	10.0	8.0	9.0		8.0	9.0	8.0	6.0	3.0	1.5	1.0	1.0	3.0	6.0	6.0	9.0
6	---	2.0	0.0	---	10.5	4.0	6.0		7.0	10.0	9.0	5.0	4.0	3.0	2.0	0.0	2.0	4.0	4.0	7.0
7	---	1.0	—0.5	---	9.0	1.0	4.0	Observations irregular.	8.0	10.0	10.0	8.0	6.0	4.0	3.0	0.0	1.0	2.0	3.0	3.0
8	---	0.0	—0.5	---	5.0	0.0	3.0		9.0	10.4	10.5	11.0	8.0	7.0	5.0	4.0	2.0	0.0	0.0	0.0
9		11.0	5.0	4.9	2.0	1.5	1.5	4.0	10.0	10.0	10.5	11.5	10.0	9.5	7.5	---	5.0	5.0	1.0	1.0
10		13.0	8.0	6.0	1.0	3.0	3.0	5.0	9.0	10.0	12.0	12.0	12.0	12.0	11.5	---	8.0	8.0	4.0	4.0
11		14.5	11.0	10.0	7.0	7.0	5.6	5.4		8.0	10.4	11.0	12.6	13.0	14.0	10.0	---	9.0	11.0	7.0	6.0
Midn't		14.2	13.0	12.0	10.0	8.6	7.0	7.0		6.6	9.0	10.0	11.6	11.0	13.0	9.0	---	10.0	12.0	9.5	8.0

Sept. 1854. | October, 1854.

Mean solar hour.	28th.	29th.	30th.	1st.	4th.	5th.	6th.	7th.	8th.	9th.	10th.	11th.	12th.	15th.	17th.	18th.	19th.	20th.	21st.	22d.
1	9.0	8.0	---	9.0	---	6.0	10.0	12.0	14.0	8.0	13.0	10.0	8.5	6.0	---	5.0	---	6.0	5.0	---
2	12.0	10.0	---	8.0	---	4.0	7.0	10.0	12.2	7.0	12.0	---	14.5	7.0	---	4.0	---	4.0	4.0	---
3	13.0	9.0	8.0	6.0	---	6.0	5.0	7.0	7.0	4.0	10.0	9.0	10.0	8.0	3.0	3.0	1.0	3.0	0.5	1.0
4	10.0	8.0	9.0		---	3.0	3.0	4.0	3.0	2.0	7.0	8.0	7.0	10.0	4.0	4.0	0.0	2.0	0.0	0.0
5	9.0	8.0	8.0		---	—1.0	—1.0	2.0	2.0	1.0	4.0	4.0	5.0	9.0	5.0	6.0	2.0	1.0	1.0	0.5
6	5.5	7.0	7.0		---	0.0	0.4	1.0	0.0	0.0	0.0	2.0	4.0	8.0	6.0	4.3	3.0	3.0	2.0	1.5
7	3.0	6.0	6.0		4.0	4.0	4.0	1.0	0.0	0.0	1.0	1.0	3.0	7.0	7.0	7.0	6.0	5.0	3.0	1.0
8	1.0	4.0	5.0		5.0	7.0	7.0	5.0	—1.0	1.5	0.0	2.0	6.0	8.0	8.0	9.0	7.0	6.0		3.0
9	2.0	3.0	5.0		10.0	10.0	10.0	8.5	7.0	7.0	2.5	2.0	3.5	5.0	9.0	11.0	12.0	10.0	8.0	7.0
10	3.0	2.0	4.0		10.0	11.5	11.5	9.0	10.0	5.0	5.5	5.0	4.0	10.0	10.0	11.5	11.0	11.0	11.0	10.0
11	7.0	4.0	3.0		9.0	12.0	12.0	13.0	11.2	10.0	7.0	11.0	6.0	3.0	8.0	8.0	10.0	12.0	12.0	13.0
Noon	9.0	6.0	4.0		8.0	13.0	12.0	14.0	13.0	12.0	10.0	13.0	8.0	4.0	6.0	7.0	8.0	10.0	12.0	13.0
1	12.0	7.0	6.0		6.0	7.0	12.0	13.0	13.0	13.0	13.0	14.0	10.0	5.2	5.0	6.0	7.0	6.0	7.5	8.0	10.0	12.0
2	10.0	9.0	7.0		4.0	4.0	8.5	9.0	12.0	14.0	14.0	13.0	13.0	5.0	4.0	5.0	4.0	3.0	7.0	9.0
3	11.0	10.0	8.0		2.0	2.0	7.0	5.0	8.0	12.5	12.0	14.0	6.0	3.0	2.0	1.0	0.5	4.0	4.0
4	12.0	11.5	9.0		0.0	0.0	3.0	3.0	3.0	7.0	9.0	10.0	14.0	7.0	4.0	2.0	1.0	0.0	2.0	1.5
5	10.0	11.0	16.0		0.0	—2.0	1.0	0.0	2.0	5.0	8.0	7.0	11.0	8.0	5.0	2.0	2.0	1.0	0.0	0.0
6	8.0	11.0	10.0		5.0	3.0	0.0	0.0	1.0	3.0	4.0	4.0	9.0	9.0	7.0	4.0	6.0	3.0	1.0	1.0
7	6.0	9.0	9.0		9.0	7.0	—0.5	2.0	0.0	3.0	4.0	3.0	7.4	9.0	7.0	7.0	7.0	4.0	1.0	3.0
8	4.0	6.0	8.0		10.0	9.0	0.0	5.0	1.0	0.0	0.5	2.0	6.0	10.0	7.0	9.5	8.0	7.0	3.0	7.0
9	2.0	3.0	7.0		12.0	13.0	9.0	7.0	4.0	3.0	3.0	4.0	8.0	8.0	12.0	9.0	8.0	7.0	10.0
10	3.0	4.0	5.5		13.0	13.2	12.0	10.0	5.0	4.0	4.0	6.0	6.0	11.0	10.0	10.0	9.0	10.0	12.0
11	5.0	5.5	5.0		14.0	14.0	11.0	13.6	9.0	8.0	7.0	7.0	8.0	5.0	8.0	7.0	10.0	9.0	12.0	13.0
Midn't	6.0	5.5	4.5		13.0	14.0	13.6	14.0	10.0	11.0	10.0	10.0	10.0	7.0	7.0	6.0	11.0	11.0	11.0	11.0

Note.—The above numbers were taken from the record, converting the fathoms into feet and deducting 8, in order to reduce the mean level of 15 feet to the adopted level of comparison of 7 feet. The observations are taken with the sounding line; bottom weedy.

Sept. 8. Some doubt about the time between 1 and 5 P. M.

After October 22 the soundings are too irregular, and later observations with the pulley-gauge too much affected by changes of the index.

This last series is considerably inferior in accuracy to the three preceding series.

Reduction of Tides, Van Rensselaer Harbor, 1853–'54.

Having given the tidal record in a form ready for use, the observations next require to be properly tabulated for the purpose of deducing empirically their laws, and for comparison with theory. In the United States Coast Survey two blank forms are in use for this tabulation; they have in their essential part been adopted as suitable for the Van Rensselaer Harbor tides, and were used with permission of the Superintendent of the Survey. They are strictly applicable only for such cases where the diurnal inequality is comparatively small, or is at least not approximating to the production of single day tides. In order to show, at a glance, the general character of the tides under discussion, they were plotted a second time, and are given in Plates I, II, and III; the observations having previously been referred to the same mean level. From these diagrams it appears that the diurnal inequality is not of so great an effect as to render the use of the ordinary method of reduction unavailable; on the other hand, it is sufficiently large to require a special discussion for time and height. The extension of the series of observations over a whole year must be considered as a fortunate circumstance, since the results thereby gain considerably in accuracy over others deduced only from a few disconnected lunations.

The tidal record would not be complete without the observations for direction and force of the wind, and for atmospheric pressure; the reader will find these records in my discussion of the meteorological material of the expedition, in Vol. XI, Smithsonian Contributions to Knowledge, 1859.

The following pages contain the first tabulation of the preceding record, viz: column 1 contains the date, civil reckoning, adopted for convenience sake. Column 2 gives the apparent time (civil reckoning) of the moon's superior and inferior transit over the Van Rensselaer meridian, obtained by adding nine minutes to the time of transit at Greenwich, allowing for a difference of longitude of $4^h\ 43\frac{1}{4}^m$ W. The mean time was converted into apparent time by applying the equation of time. The time for the lower transit was obtained by taking the mean of the time of the preceding and following upper transit. Columns 3 and 4 contain the apparent time of high and low water, taken from the record; in some cases a graphical method was resorted to, to obtain the instant of these phases with greater precision. The equation of time has been applied to the mean time in which the observations are expressed. Columns 5 and 6 contain the lunitidal interval between the time of high water and low water, and the time of the transit of the moon immediately preceding, though in some cases, owing to the half-monthly inequality, it may be the second preceding, the establishment being about $11\frac{3}{4}$ hours. This transit of comparison has been called transit F by Mr. Lubbock.[1] The next columns, 7 and 8, give the height of high and low water, extracted from the preceding abstract. The remaining columns contain the moon's parallax and declination at noon.

---

[1] See an Elementary Treatise on the Tides, by J. W. Lubbock, Esq., London, 1839.

TABLE FOR THE REDUCTION OF TIDES.—No. 1.

Showing the times of High and Low Water, and the Heights of High and Low Tides; together with the time of the Moon's passing the Meridian of the place, and the Lunitidal Intervals, at Van Rensselaer Harbor during the months of October 10, 1853, to October 22, 1855.

SERIES I.—FROM OCTOBER 10 TO DECEMBER 28, 1853.

Date. 1853.	Moon passes the meridian. Appar. time.		Apparent time of				Lunitidal interval.				Height of				Moon's parallax at noon.		Moon's declination at noon.	
			H. water.		L. water.		H. water.		L. water.		H. water.		L. water.					
	H.	M.	H.	M.	H.	M.	H.	M.	H.	M.	Ft.	Dec.	Ft.	Dec.	Min.	Dec.	Degree.	Dec.
Oct. 9	6	28	...	...	...	...	...	...	...	...	...	...	...	...	...	...	...	...
" 10	6	57	8	13	11	13	13	45	16	45	8	0	4	7	58	4	—23	8
	7	26	6	13	...	...	11	16	...	...	9	2	...	...				
" 11	7	54	7	58	1	43	12	32	18	46	6	7	4	4	57	8	—20	9
	8	22	7	43	1	13	11	49	17	47	9	5	4	8				
" 12	8	47	8	13	1	43	11	51	17	49	7	9	4	0	57	3	—17	0
	9	12	9	29	2	43	12	42	18	21	10	0	4	2				
" 13	9	37	9	59	3	14	12	47	18	27	9	1	3	7	56	7	—12	3
	10	02	10	14	3	59	12	37	18	47	10	9	3	1				
" 14	10	24	10	14	4	14	12	12	18	37	9	8	2	7	56	3	— 7	1
	10	47	10	44	3	59	12	20	17	57	11	3	2	7				
" 17	0	13	...	...	...	...	...	...	...	...	...	...	...	...	54	9	+ 8	9
	0	35	11	45	5	15	11	10	17	02	10	7	1	5				
" 18	0	57	...	...	6	00	...	...	17	25	...	...	1	8	54	5	+13	7
	1	18	...	...	6	30	...	...	17	33	12	3	2	0				
" 19	1	40	0	30	6	45	11	12	17	27	11	5	1	6	54	3	+17	9
	2	02	1	15	7	15	11	35	17	35	12	4	1	9				
" 20	2	25	1	15	5	45	11	13	15	43	11	5	5	7	54	2	+21	3
	2	48	1	45	9	00	11	20	18	35	11	9	2	7				
" 21	3	12	1	30	7	45	10	42	16	57	9	9	2	8	54	1	+23	8
	3	36	3	15	8	30	12	03	17	18	11	1	4	4				
" 22	4	00	1	15	9	16	9	39	17	40	10	5	3	4	54	2	+25	2
	4	25	3	16	8	46	11	16	16	46	9	9	4	5				
" 23	4	51	3	16	8	31	10	51	16	06	10	3	3	9	54	4	+25	5
	5	16	4	16	10	16	11	25	17	25	10	4	5	2				
" 24	5	42	3	31	8	31	10	15	15	15	8	7	4	5	54	8	+24	7
	6	07	5	46	...	...	12	04	...	...	9	7	...	...				
" 25	6	32	5	01	0	16	10	54	18	34	7	1	5	4	55	4	+22	6
	6	57	8	16	10	46	1	44	16	39	8	5	5	2				
" 26	7	22	7	46	0	31	12	49	17	59	7	0	5	2	56	1	+19	5
	7	46	8	46	0	46	13	24	17	49	9	2	6	1				
" 27	8	11	8	16	1	46	12	30	18	24	8	1	4	8	57	0	+15	3
	8	35	8	46	0	46	12	35	17	00	9	8	5	5				
" 28	6	59	...	...	3	01	...	...	18	50	9	3	4	5	57	9	+10	3
	9	23	9	46	3	46	12	47	19	11	10	8	5	1				
" 29	9	47	9	46	4	01	12	23	19	02	10	7	3	4	58	8	+ 4	6
	10	11	10	46	4	16	12	59	18	53	11	4	4	9				
" 30	10	36	12	01	3	46	13	50	17	59	11	6	2	9	59	7	— 1	5
	11	01	10	46	5	16	12	10	19	05	11	7	3	9				
" 31	11	26	11	31	4	16	12	30	17	40	12	0	1	6	60	4	— 7	7
	11	52	12	16	5	01	12	50	18	00	12	2	1	8				
Nov. 1	...	...	11	46	4	46	11	54	17	20	12	3	2	7	60	8	—13	6
	0	19	11	16	7	01	10	57	19	09	11	6	3	6				
" 2	0	48	...	...	7	01	...	...	18	42	...	...	0	9	61	0	—18	7
	1	16	0	16	7	16	11	28	18	28	14	3	0	1				
" 3	1	46	1	16	7	16	12	00	18	00	11	8	0	0	60	9	—22	7
	2	16	1	31	8	46	11	45	19	00	14	2	1	5				
" 4	2	47	1	16	8	01	11	00	17	45	11	6	0	3	60	5	—25	0
	3	19	...	...	8	01	...	...	17	14	...	...	1	6				
" 5	3	50	1	16	8	31	9	57	17	12	10	1	0	7	59	9	—25	6
	4	22	2	46	8	46	10	56	16	56	13	0	1	8				
" 6	4	52	4	01	9	01	11	39	16	39	9	1	2	3	59	1	—24	5
	5	23	4	46	10	31	11	54	17	39	10	9	2	6				
" 7	5	52	3	46	10	16	10	23	16	53	8	2	3	2	58	4	—21	9
	6	21	4	31	...	...	10	39	...	...	10	7	...	...				

12   RECORD AND REDUCTION OF THE TIDES.

SERIES I.—FROM OCTOBER 10 TO DECEMBER 28, 1853.

Date 1853.	Moon passes the meridian. Appar. time. H. M.	Apparent time of		Lunitidal interval.		Height of		Moon's parallax at noon.		Moon's declination at noon.	
		H. water. H. M.	L. water. H. M.	H. water. H. M.	L. water. H. M.	L. water. Ft. Dec.	H. water. Ft. Dec.	Min.	Dec.	Degree.	Dec.
Nov. 8	6 48	5 16	6 16	10 55	18 24	7 8	3 5	57	6	—18	1
	7 14	6 46	0 16	11 58	17 55	10 4	4 9				
" 9	7 38	7 16	0 46	12 02	17 58	8 1	4 0	56	9	—13	5
	8 02	7 31	1 16	11 53	18 02	11 0	4 4				
" 10	8 25	8 46	0 46	12 44	17 08	9 5	4 3	56	3	— 8	4
	8 48	9 46	2 31	13 21	18 29	10 7	5 5				
" 11	9 09	10 01	2 46	13 13	18 21	9 5	2 3	55	7	— 3	1
	9 31	9 31	3 31	12 22	18 43	11 8	4 2				
" 12	9 52	10 01	4 16	12 30	19 07	11 2	2 8	55	2	+ 2	3
	10 13	9 46	3 31	11 54	18 00	11 5	3 8				
" 13	10 34	10 16	4 46	12 03	18 54	11 5	2 6	54	8	+ 7	5
	10 54	11 15	6 15	12 41	20 02	11 0	2 5				
" 14	11 15	10 45	5 0	11 51	18 26	12 7	2 1	54	5	+12	4
	11 37	10 45	5 30	11 30	18 30	9 8	2 9				
" 15	11 59	11 30	5 30	11 53	18 15	13 3	0 7	54	2	+16	8
	... ...	12 00	6 30	12 01	18 53	13 0	3 5				
" 16	0 21	12 30	5 0	12 09	17 01	13 6	0 4	54	1	+20	5
	0 44	11 45	7 15	11 01	18 54	10 9	3 1				
" 17	1 07	... ...	7 0	... ...	18 16	... ...	1 3	54	0	+23	2
	1 31	0 30	0 45	11 25	17 38	13 0	3 1				
" 18	1 55	0 45	7 15	11 14	17 44	13 7	—0 1	54	0	+25	0
	2 20	1 15	8 15	11 20	18 20	12 1	1 2				
" 19	2 44	1 14	7 14	10 54	16 54	9 8	1 9	54	1	+25	7
	3 09	1 44	8 14	11 00	17 30	13 0	2 3				
" 20	3 34	1 44	8 14	10 35	17 05	9 1	2 6	54	4	+25	2
	3 59	2 14	8 59	10 40	17 25	11 5	3 6				
" 21	4 24	2 29	7 59	10 30	16 00	8 0	3 2	54	7	+23	5
	4 49	2 44	11 14	10 20	18 50	10 7	3 6				
" 22	5 14	2 49	8 44	10 00	15 55	7 4	5 1	55	3	+20	7
	5 37	3 43	11 13	10 29	17 59	10 1	3 7				
" 23	6 01	6 13	10 28	12 36	16 51	7 1	6 6	55	9	+16	9
	6 25	4 43	... ...	10 42	... ...	9 9	... ...				
" 24	6 48	8 13	0 43	13 48	18 32	8 5	3 7	56	6	+12	3
	7 11	6 43	0 13	11 55	17 48	9 5	3 4				
" 25	7 34	7 48	0 43	12 37	17 55	8 8	3 7	57	7	+ 7	0
	7 56	6 57	1 42	11 23	18 31	10 3	6 9				
" 26	8 20	8 42	1 12	12 46	17 38	9 4	4 2	58	6	+ 1	2
	8 44	8 27	2 42	12 07	18 46	9 0	4 3				
" 27	9 08	9 42	3 42	12 58	19 22	10 9	2 7	59	6	— 4	9
	9 33	9 12	4 12	12 04	19 28	11 2	3 2				
" 28	9 58	10 12	4 12	12 30	19 04	12 0	2 2	60	5	—10	9
	10 24	11 11	4 41	13 13	19 08	12 5	2 6				
" 29	10 51	11 41	4 41	13 17	18 43	13 8	0 7	61	0	—16	5
	11 20	10 56	4 11	12 05	17 47	10 7	1 2				
" 30	11 50	10 56	4 41	11 36	17 50	13 7	0 1	61	4	—21	1
	... ...	... ...	5 26	... ...	18 06	... ...	1 7				
Dec. 1	0 21	0 11	5 41	12 21	17 51	10 8	—1 2	61	4	—24	3
	0 53	0 11	6 26	11 50	18 05	12 8	—0 3				
" 2	1 25	0 10	6 10	11 17	17 17	9 5	—2 1	61	0	—25	7
	1 58	1 10	7 10	11 45	17 45	13 6	0 0				
" 3	2 30	1 10	7 10	11 12	17 12	9 4	—0 4	60	4	—25	2
	3 03	1 55	8 40	11 25	18 10	12 5	—0 1				
" 4	3 32	1 54	6 39	10 51	15 36	8 3	0 1	59	6	—22	9
	4 03	2 09	9 39	10 37	17 37	13 7	1 6				
" 5	4 31	2 39	8 54	10 36	16 51	9 1	1 3	58	7	—19	4
	5 00	3 39	9 39	11 08	17 08	11 1	1 0				
" 6	5 26	3 09	8 24	10 09	15 24	6 1	2 3	57	7	—14	9
	5 52	4 24	11 39	10 58	18 13	10 3	1 0				
" 7	6 14	6 38	... ...	12 46	... ...	6 8	... ...	56	9	— 9	8
	6 38	7 53	0 08	13 39	18 16	10 6	3 4				
" 8	7 00	6 23	1 08	11 45	18 54	7 6	3 5	56	1	— 4	4
	7 22	... ...	0 38	... ...	18 00	... ...	2 9				
" 9	7 42	8 37	1 07	13 15	18 07	11 2	3 6	55	4	+ 1	0
	8 04	8 07	2 07	12 25	18 45	11 8	5 2				
" 10	8 25	9 37	2 07	13 33	18 25	11 1	5 3	54	9	+ 6	3
	8 46	9 07	2 07	12 42	18 03	9 2	5 3				

RECORD AND REDUCTION OF THE TIDES. 43

Date. 1853.	Moon passes the meridian. Appar. time.		Apparent time of				Lunitidal interval.				Height of				Moon's parallax at noon.		Moon's declination at noon.	
			H. water.		L. water.		H. water.		L. water.		H. water.		L. water.					
	H.	M.	H.	M.	H.	M.	H.	M.	H.	M.	Ft.	Dec.	Ft.	Dec.	Min.	Dec.	Degree.	Dec.

Series I.—From October 10 to December 28, 1853.

Date 1853																		
Dec. 11	9	06	9	51	3	36	13	05	19	11	9	5	1	7	54	5	+11	2
	9	27	9	36	3	36	12	30	18	50	7	8	1	8				
" 12	9	49	11	06	4	06	13	39	19	00	9	9	0	3	54	2	+15	7
	10	10	11	06	4	36	13	17	19	09	9	6	2	8				
" 13	10	31	10	35	4	05	12	35	18	16	11	0	3	5	54	0	+19	6
	10	54	10	20	4	35	11	49	18	25	8	6	2	8				
" 14	11	17	11	05	4	35	12	11	18	04	12	5	2	3	53	9	+22	6
	11	41	11	05	0	05	11	48	19	11	8	9	2	4				
" 15	...	...	...	...	5	05	...	...	17	48	...	...	2	4	53	9	+24	7
	0	06	1	04	6	04	13	23	18	23	13	2	5	5				
" 16	0	30	0	04	5	04	11	58	16	58	12	5	3	4	54	0	+25	6
	0	55	0	19	7	19	11	49	18	49	12	8	1	6				
" 17	1	20	0	34	6	03	11	39	17	08	7	5	2	7	54	2	+25	4
	1	44	...	...	5	18	...	...	15	58	...	...	0	6				
" 18	2	09	1	03	7	03	11	19	17	19	...	...	...	...	54	5	+24	0
	2	34	...	...	...	...	...	...	...	...	...	...	...	...				
" 19	2	58	2	03	7	32	11	29	17	58	...	...	...	...	54	8	+21	5
	3	22	1	32	8	02	10	34	17	04	...	...	...	...				
" 20	3	46	2	17	8	32	10	55	17	10	10	9	1	4	55	3	+18	0
	4	09	2	47	6	02	11	01	14	16	10	4	...	...				
" 21	4	32	3	02	8	31	10	53	16	22	11	6	4	5	55	9	+13	7
	4	54	3	46	9	31	11	14	16	59	11	5	4	4				
" 22	5	17	4	31	9	16	11	37	16	22	9	1	5	1	56	5	+ 8	7
	5	39	4	31	11	01	11	14	17	44	11	6	4	2				
" 23	6	01	5	31	11	30	11	52	17	51	7	5	3	1	57	4	+ 3	2
	6	24	4	30	11	30	10	29	17	29	8	8	2	9				
" 24	6	47	7	00	...	...	12	36	...	...	7	8	...	...	58	3	− 2	6
	7	10	6	30	0	00	11	43	17	36	9	3	4	6				
" 25	7	34	8	00	1	15	12	50	18	28	8	0	2	4	59	2	− 8	5
	7	59	7	30	0	15	11	56	17	05	7	3	4	1				
" 26	8	23	8	29	1	44	12	30	18	10	11	5	3	2	60	0	−14	1
	8	50	8	14	2	44	11	51	18	45	9	2	4	8				
" 27	9	18	9	28	1	59	12	38	17	36	11	2	1	8	60	7	−19	1
	9	46	9	28	3	58	12	10	19	08	9	1	2	9				
" 28	10	16	10	13	3	28	12	27	18	10	11	6	1	9	61	2	−22	9
	10	48	10	58	4	58	12	42	19	12	10	6	4	9				

Series II.—From January 28 to April 7, 1854.

Jan. 27	10	58	...	...	...	...	...	...	...	...	...	...	...	...	60	7	−25	0
	11	30	...	...	...	...	...	...	...	...	...	...	...	...				
" 28	...	...	0	17	5	17	13	19	18	19	10	6	1	0	60	4	−22	5
	0	01	0	17	6	17	12	47	18	47	12	4	2	6				
" 29	0	31	0	17	6	17	12	16	18	16	11	9	1	3	59	9	−18	6
	1	01	0	2	8	02	11	31	19	31	13	3	−1	8				
" 30	1	27	0	16	6	01	11	15	17	00	9	0	0	5	59	1	−13	7
	1	55	1	16	7	46	11	49	18	19	13	9	−0	4				
" 31	2	20	0	46	7	16	10	51	17	21	9	7	−0	3	58	2	− 8	1
	2	45	1	16	8	16	10	56	17	56	13	7	1	5				
Feb. 1	3	08	2	46	8	46	12	01	18	01	11	1	1	3	57	3	− 2	4
	3	32	2	46	9	01	11	38	17	53	12	9	2	1				
" 2	3	54	4	01	9	01	12	20	17	29	10	0	2	0	56	4	+ 3	3
	4	16	2	46	10	01	10	52	18	07	10	6	0	3				
" 3	4	38	4	01	8	46	11	45	16	30	8	5	3	4	55	6	+ 8	7
	5	00	3	46	9	31	11	08	16	53	8	6	2	2				
" 4	5	21	4	01	9	01	11	46	16	01	9	2	3	8	55	0	+13	6
	5	43	5	16	12	16	11	55	18	55	8	9	3	8				
" 5	6	06	6	16	12	16	12	33	18	33	10	5	5	0	54	5	+17	8
	6	28	5	16	11	46	11	10	17	40	10	7	4	5				
" 6	6	51	8	16	11	46	13	48	17	18	9	5	3	8	54	3	+21	3
	7	14	8	01	11	16	13	10	16	25	7	2	3	7				
" 7	7	38	7	46	...	...	12	32	...	...	9	4	...	...	54	1	+23	9
	8	02	7	31	3	01	11	53	19	47	8	2	5	8				

Series II.—From January 28 to April 7, 1854.

Date. 1854.	Moon passes the meridian. Appar. time.		Apparent time of				Lunitidal interval.				Height of				Moon's parallax at noon.		Moon's declination at noon.	
			H. water.		L. water.		H. water.		L. water.		H. water.		L. water.					
	H.	M.	H.	M.	H.	M.	H.	M.	H.	M.	Ft.	Dec.	Ft.	Dec.	Min.	Dec.	Degree.	Dec.
Feb. 8	8	26	10	01	2	46	13	50	19	08	9	5	4	7	54	1	+25	4
	8	50	9	15	3	15	12	49	19	13	6	7	3	3				
" 9	9	15	10	15	3	15	13	25	18	49	9	7	3	8	54	3	+25	7
	9	41	9	45	4	45	12	30	19	55	7	2	3	3				
" 10	10	06	10	15	6	15	12	34	21	00	9	2	3	7	54	6	+24	9
	10	32	11	45	5	15	13	39	19	31	6	6	1	5				
" 11	10	57	11	45	5	00	13	13	18	54	10	5	1	3	54	9	+22	8
	11	22	11	45	4	45	12	48	18	13	7	7	1	8				
" 12	11	46	10	45	5	00	11	23	18	03	11	5	2	8	55	4	+19	8
	...	...	11	15	6	30	11	29	19	08	8	4	2	3				
" 13	0	10	11	15	5	45	11	05	17	59	12	6	0	7	55	8	+15	6
	0	33	10	45	5	15	10	12	17	05	10	6	4	1				
" 14	0	58	...	...	6	46	...	...	18	13	...	...	2	4	56	3	+10	7
	1	21	0	46	7	16	11	48	18	18	13	6	2	6				
" 15	1	44	0	31	6	46	11	10	17	25	11	8	3	2	56	8	+ 5	3
	2	06	1	16	7	46	11	32	18	02	12	9	4	1				
" 16	2	29	0	16	7	46	10	10	17	40	13	1	2	2	57	3	− 0	3
	2	51	2	01	8	16	11	32	17	47	13	1	2	3				
" 17	3	14	3	16	7	16	12	25	16	25	12	6	3	9	57	7	− 6	0
	3	37	...	...	9	16	...	...	18	02	11	7	0	3				
" 18	4	00	3	01	9	01	11	24	17	24	10	6	2	7	58	2	−11	7
	4	24	2	16	9	16	10	16	17	16	10	2	3	0				
" 19	4	49	3	16	10	46	10	52	18	22	11	3	4	5	58	7	−16	8
	5	16	3	16	10	16	10	27	17	27	10	7	3	0				
" 20	5	42	3	46	9	16	10	30	16	00	10	8	5	5	59	1	−20	9
	6	10	3	46	...	...	10	04	...	...	10	7	...	...				
" 21	6	39	...	...	...	...	...	...	...	...	...	...	...	...	59	5	−24	1
	7	09	...	...	11	46	...	...	17	36	...	...	4	3				
" 22	7	40	...	...	...	...	...	...	...	...	...	...	...	...	59	8	−25	7
	8	10	7	46	4	16	12	06	21	07	9	1	4	7				
" 23	8	41	10	16	2	46	14	06	19	06	10	7	4	1	59	9	−25	6
	9	12	8	41	4	46	12	05	20	36	8	9	5	7				
" 24	9	45	9	17	1	17	12	05	16	36	10	3	3	3	59	9	−23	7
	10	15	10	17	3	47	12	32	18	35	9	7	2	8				
" 25	10	45	10	17	3	47	12	02	17	52	11	2	2	8	59	6	−20	3
	11	13	10	47	4	17	12	02	18	02	10	4	2	0				
" 26	11	41	12	47	4	47	13	34	18	02	11	9	1	0	59	2	−15	7
	...	...	11	47	6	02	12	06	18	49	11	3	3	0				
" 27	0	07	11	47	5	02	11	40	17	21	12	7	−0	4	58	6	−10	3
	0	33	11	47	6	02	11	14	17	55	10	3	−0	9				
" 28	0	57	12	32	6	17	11	35	17	44	13	0	0	1	57	9	− 4	5
	1	22	11	47	7	02	10	25	18	05	12	4	1	0				
March 1	1	45	...	...	7	02	...	...	17	40	...	...	0	5	57	1	+ 1	4
	2	09	1	32	8	47	11	47	19	02	12	9	1	3				
" 2	2	31	1	18	7	48	11	09	17	39	12	0	2	5	56	4	+ 7	0
	2	53	1	33	9	18	11	02	18	47	13	0	1	9				
" 3	3	15	2	03	8	18	11	10	17	25	10	6	2	5	55	7	+12	2
	3	38	2	18	9	03	11	03	17	48	10	9	1	5				
" 4	4	00	3	03	7	48	11	25	16	10	11	1	3	0	55	0	+16	7
	4	23	2	48	8	48	10	48	16	48	11	1	1	0				
" 5	4	46	3	48	8	48	11	23	16	25	9	2	3	1	54	6	+20	5
	5	09	3	03	9	48	10	17	17	02	10	1	3	9				
" 6	5	32	4	18	10	03	11	09	16	54	10	5	6	1	54	3	+23	4
	5	56	5	19	10	34	11	47	17	02	9	6	4	1				
" 7	6	22	4	49	10	49	10	53	16	53	8	8	5	8	54	2	+25	3
	6	47	...	...	...	...	...	...	...	...	...	...	...	...				
" 9	8	03	...	...	...	...	...	...	...	...	...	...	...	...			+25	5
	8	28	...	...	...	...	...	...	...	...	...	...	...	...				
" 10	8	53	9	49	4	21	13	21	20	18	9	9	3	9	54	9	+23	8
	9	18	10	50	4	35	13	57	20	07	8	8	5	0				
" 11	9	42	10	50	4	35	13	32	19	42	10	6	5	4	55	4	+21	0
	10	08	...	...	...	...	...	...	...	...	...	...	...	...				
" 14	11	43	...	...	...	...	...	...	...	...	...	...	...	...			+ 7	1
	0	06	12	21	...	...	12	15	...	...	11	2	...	...				

RECORD AND REDUCTION OF THE TIDES. 45

SERIES II.—FROM JANUARY 28 TO APRIL 7, 1854.

DATE. 1854.	Moon passes the meridian. Appar. time.		Apparent time of				Lunitidal interval.				Height of				Moon's parallax at noon.		Moon's declination at noon.	
			H. water.		L. water.		H. water.		L. water.		H. water.		L. water.					
	H.	M.	H.	M.	H.	M.	H.	M.	H.	M.	Ft.	Dec.	Ft.	Dec.	Min.	Dec.	Degree.	Dec.
Mar. 15	0	29	11	21	6	21	10	52	18	15	12	7	2	3	57	6	+ 1	2
	0	52	...	...	8	06	...	...	19	37	...	...	2	7				
" 16	1	15	0	36	8	21	11	44	19	29	10	5	3	2	58	1	— 4	6
	1	38	1	21	6	34	12	06	17	21	8	4	3	0				
" 17	2	02	1	36	8	21	11	58	18	43	12	0	0	7	58	5	—10	5
	2	26	...	...	8	21	...	...	18	19	10	4	1	2				
" 18	2	52	1	52	9	22	11	26	18	56	12	1	1	6	58	8	—15	8
	3	18	1	52	8	07	11	00	17	15	9	5	3	1				
" 19	3	44	2	22	7	52	11	04	16	34	12	3	2	7	59	0	—20	4
	4	12	2	37	8	37	10	53	16	53	10	7	2	6				
" 20	4	40	4	22	...	...	12	10	...	...	12	4	...	...	59	2	—23	8
" 22	7	11	...	...	...	...	...	...	...	...	...	...	...	...			—26	0
" 23	7	42	7	23	...	...	12	12	...	...	9	5	...	...	59	2	—24	6
	8	12	8	23	1	53	12	41	18	42	8	5	5	3				
" 24	8	42	7	54	2	09	11	42	18	27	10	2	4	8	59	0	—21	6
	9	11	10	39	2	24	13	57	18	12	9	2	5	3				
" 25	9	39	9	54	3	54	12	43	19	12	10	3	5	1	58	7	—17	5
	10	05	9	54	3	54	12	15	18	43	9	6	4	3				
" 26	10	31	9	24	4	24	11	19	18	45	11	7	3	8	58	3	—12	4
	10	55	11	24	4	39	12	53	18	34	11	4	3	4				
" 27	11	19	10	54	4	54	11	59	18	23	11	9	2	2	57	8	— 6	7
	11	43	11	25	5	25	12	06	18	30	10	9	1	3				
" 28	...	...	11	25	5	55	11	42	18	36	12	0	2	0	57	3	— 0	8
	0	07	...	...	6	25	...	...	18	42	...	...	—1	3				
" 29	0	29	0	40	6	25	12	33	18	18	12	0	3	5	56	7	+ 4	9
	0	52	0	40	6	25	12	11	17	56	14	5	—0	2				
" 30	1	14	1	55	7	55	13	03	19	03	12	5	3	3	56	1	+10	4
	1	36	0	0	7	41	10	46	18	27	11	5	1	3				
" 31	1	59	1	41	7	56	12	05	18	20	11	5	1	5	55	5	+15	3
	2	22	1	56	8	56	11	57	18	57	13	0	3	3				
April 4	5	11	...	...	...	...	...	...	...	...	...	...	...	...	54	3	+26	1
	5	36	3	57	10	12	10	46	17	01	9	3	4	2				
" 5	6	01	5	42	10	42	12	06	17	06	9	7	5	0	54	3	+26	1
	6	27	5	42	10	42	11	41	16	41	6	9	5	7				
" 6	6	52	6	28	...	...	12	01	...	...	8	8	...	...	54	6	+24	8
	7	17	8	58	2	28	14	06	20	01	9	7	6	0				
" 7	7	41	8	13	0	0	12	56	17	08	8	5	5	5	55	0	+22	3
	8	07	9	58	2	28	14	17	19	11	7	5	5	5				

SERIES III.—FROM APRIL 20 TO AUGUST 3, 1854.

April 19	6	17	...	...	...	...	...	...	...	...	...	...	...	...				
" 20	6	45	...	...	...	...	...	...	...	...	...	...	...	...	58	8	—22	6
	7	13	6	01	0	16	11	16	17	59	8	1	5	1				
" 21	7	41	7	31	0	31	12	18	17	46	10	0	4	5	58	4	—18	7
	8	07	7	46	1	31	12	05	18	18	7	2	4	0				
" 22	8	35	8	02	1	17	11	55	17	36	10	4	4	0	57	9	—13	8
	8	59	9	47	3	17	13	12	19	10	9	8	4	0				
" 23	9	24	8	17	3	02	11	18	18	27	10	7	4	1	57	4	— 8	5
	9	47	19	17	4	32	12	53	19	33	9	6	1	8				
" 24	10	10	9	02	3	32	11	15	18	08	11	1	2	5	57	0	— 2	6
	10	32	9	32	4	32	11	22	18	45	9	9	1	7				
" 25	10	54	...	02	5	32	11	30	19	22	12	2	1	6	56	4	+ 3	1
	11	16	...	...	5	04	...	...	18	32	...	...	1	8				
" 26	11	38	...	...	4	02	...	...	17	08	...	...	2	6	56	0	+ 8	7
	12	00	0	02	6	32	12	46	19	16	12	0	0	1				
" 27	...	...	1	02	6	47	13	24	19	09	11	5	1	0	55	5	+13	8
	0	22	0	32	5	02	12	32	17	02	12	2	2	2				
" 28	0	45	...	...	5	03	...	...	16	41	...	...	2	0	55	0	+18	3
	1	09	1	03	7	03	12	18	18	18	11	6	2	2				
" 29	1	33	0	33	7	33	11	24	18	24	13	0	2	1	54	7	+21	9
	1	56	0	03	5	33	10	30	16	00	12	0	1	5				

SERIES III.—FROM APRIL 20 TO AUGUST 3, 1854.

Date. 1854.	Moon passes the meridian. Appar. time.		Apparent time of				Lunitidal interval.				Height of				Moon's parallax at noon.		Moon's declination at noon.	
			H. water.		L. water.		H. water.		L. water.		H. water.		L. water.					
	H.	M.	H.	M.	H.	M.	H.	M.	H.	M.	Ft.	Dec.	Ft.	Dec.	Min.	Dec.	Degree.	Dec.
April 30	2	20	0	03	5	33	10	07	15	37	12	9	1	9	54	4	+24	5
	2	45	0	33	5	17	10	13	14	57	11	3	1	7				
May 1	3	09	1	18	9	03	10	33	18	18	12	6	3	9	54	2	+26	0
	3	34	0	33	7	03	9	24	15	54	9	4	3	7				
" 2	3	59	2	33	9	03	10	59	17	29	12	4	1	8	54	2	+26	3
	4	25	2	03	8	33	10	34	16	34	10	3	4	3				
" 3	4	50	3	48	9	03	11	23	16	38	9	6	5	3	54	3	+25	4
	5	15	3	03	9	03	10	13	16	13	9	1	6	5				
" 4	5	39	4	33	...	...	11	18	...	...	10	5	...	...	54	6	+23	4
	6	04	...	...	...	...	...	...	...	...	...	...	...	...				
" 6	7	14	...	...	...	...	...	...	...	...	...	...	...	...	55	6	+16	3
	7	38	...	...	...	...	...	...	...	...	...	...	...	...				
" 7	8	02	6	34	0	49	10	56	17	35	8	9	6	0	56	4	+11	4
	8	25	9	34	2	04	13	32	18	26	7	9	3	6				
" 8	8	48	8	34	1	04	12	09	17	02	11	5	5	2	57	3	+ 6	0
	9	10	10	19	4	04	13	31	19	39	8	1	2	7				
" 9	9	33	8	34	2	04	11	24	17	19	11	3	4	4	58	1	0	0
	9	56	9	04	3	19	11	31	18	09	9	4	1	8				
" 10	10	19	10	49	5	34	12	53	20	01	8	9	2	0	58	5	— 6	0
	10	43	12	04	3	34	13	46	17	38	11	8	2	2				
" 11	11	09	11	04	4	34	12	21	18	15	10	5	1	9	59	7	—12	0
	11	35	10	49	5	04	11	40	18	21	13	7	1	3				
" 12	12	02	12	04	5	40	12	29	18	40	11	0	1	0	60	2	—17	4
	...	...	11	34	4	34	11	32	16	59	13	1	0	9				
" 13	0	30	10	49	5	19	10	19	...	...	12	4	...	...	60	5	—22	0
	1	00	11	34	5	19	10	04	16	49	13	6	1	3				
" 14	1	30	...	...	...	...	...	...	...	...	...	...	...	...	60	6	—25	0
	2	01	0	4	6	19	10	34	16	49	11	3	0	8				
" 15	2	33	0	49	7	34	10	48	17	33	13	9	1	7	60	4	—26	3
	3	05	1	49	8	19	11	16	17	46	12	3	0	5				
" 16	3	38	0	49	8	04	9	44	16	59	12	4	1	8	60	0	—25	8
	4	09	2	34	7	19	10	56	15	41	10	4	2	3				
" 17	4	41	2	49	9	04	10	40	16	55	12	8	1	8	59	4	—23	5
	5	10	2	49	8	19	10	08	15	38	10	6	2	0				
" 18	5	40	3	49	10	49	10	39	17	39	12	5	1	8	58	6	—19	8
	6	07	3	49	9	19	10	09	15	39	9	6	4	2				
" 19	6	34	5	19	...	...	11	12	...	...	11	7	...	...	58	1	—15	0
	6	59	6	19	0	04	11	45	17	57	10	2	3	0				
" 20	7	24	7	19	0	34	12	20	18	00	9	8	5	2	57	5	— 9	7
	7	47	8	34	3	04	13	10	20	05	8	7	3	8				
" 21	8	10	7	34	2	19	11	47	18	55	11	1	4	4	56	9	— 4	0
	8	32	8	34	2	34	12	24	18	47	8	6	3	7				
" 22	8	55	8	04	1	34	11	32	17	24	10	6	3	4	56	3	+ 1	6
	9	16	9	04	3	19	12	09	18	47	9	3	2	5				
" 23	9	38	8	04	3	04	10	48	18	09	11	3	3	5	55	8	+ 7	3
	9	59	10	19	4	19	12	41	19	03	11	4	3	3				
" 24	10	21	8	48	4	48	10	49	19	10	9	9	4	0	55	3	+12	4
	10	43	11	48	4	18	13	27	18	19	12	9	2	0				
" 25	11	05	10	18	5	03	11	35	18	42	10	0	2	0	54	9	+17	1
	11	28	11	03	5	33	11	58	18	50	10	8	2	0				
" 26	11	51	12	45	5	03	13	17	17	58	9	7	2	7	54	6	+20	9
	...	...	12	45	6	33	12	54	19	05	12	5	3	4				
" 27	0	15	10	48	6	18	10	33	18	27	9	7	2	4	54	3	+23	9
	0	39	...	...	4	18	...	...	16	03	...	...	0	5				
" 28	1	04	0	33	6	18	11	54	18	09	10	7	1	6	54	1	+25	7
	1	29	0	03	4	48	10	59	15	44	9	7	2	3				
" 29	1	54	0	03	7	03	10	34	17	34	10	9	1	5	54	0	+26	3
	2	19	0	48	7	18	10	54	17	24	8	5	1	6				
" 30	2	44	1	03	8	18	10	44	17	59	12	0	2	5	54	0	+25	8
	3	09	2	18	6	48	11	34	16	04	8	5	2	5				
" 31	3	34	3	18	9	18	12	09	19	09	11	8	2	5	54	2	+24	1
	3	59	2	33	8	33	10	59	16	59	8	5	3	5				
June 1	4	22	2	18	9	47	10	19	17	48	11	0	4	0	54	5	+21	4
	4	47	4	17	10	02	11	55	17	40	8	0	3	6				

RECORD AND REDUCTION OF THE TIDES. 47

SERIES III.—FROM APRIL 20 TO AUGUST 3, 1854.

Date. 1854.	Moon passes the meridian. Appar. time.		Apparent time of				Lunitidal interval.				Height of				Moon's parallax at noon.		Moon's declination at noon.	
			H. water.		L. water.		H. water.		L. water.		H. water.		L. water.					
	H.	M.	H.	M.	H.	M.	H.	M.	H.	M.	Ft.	Dec.	Ft.	Dec.	Min.	Dec.	Degree.	Dec.
June 2	5	09	4	02	10	32	11	15	17	45	9	8	3	7	54	9	+17	6
	5	32	4	47	10	17	11	38	17	08	7	8	4	8				
" 3	5	54	...	...	...	...	...	...	...	...	9	6	4	5	55	5	+13	1
	6	16	4	02	10	02	10	08	16	08	8	1	5	2				
" 4	6	38	...	...	...	...	...	...	...	...	9	4	...	...	56	3	+ 8	0
	7	00	7	32	12	02	12	54	17	24	8	6	5	6				
" 5	7	22	8	02	...	...	13	02	...	...	9	7	...	...	57	2	+ 2	3
	7	44	9	02	0	32	13	40	17	32	9	2	4	7				
" 6	8	07	9	02	1	32	13	18	18	10	10	1	6	1	58	2	− 3	6
	8	30	8	02	2	32	11	55	18	48	10	4	4	0				
" 7	8	54	9	02	4	02	12	32	19	55	10	2	4	3	59	0	− 9	5
	9	18	9	02	3	02	12	08	18	32	11	1	3	4				
" 8	9	44	8	01	2	47	10	43	17	53	9	3	4	0	59	9	−15	2
	10	10	10	01	2	31	12	17	17	13	11	6	1	8				
" 9	10	38	...	...	...	...	...	...	...	...	...	...	...	...	60	6	−20	1
	11	07	9	01	3	01	10	23	16	51	12	9	1	4				
" 10	11	38	10	01	5	01	10	54	18	23	11	7	2	1	61	1	−23	0
	...	...	13	31	6	01	13	53	18	54	11	4	0	6				
" 11	0	10	10	31	6	01	10	21	18	23	9	4	1	8	61	2	−26	0
	0	43	13	01	6	31	12	18	18	21	13	2	2	0				
" 12	1	16	11	31	6	31	10	15	17	48	10	4	1	6	61	1	−26	2
	1	49	...	...	6	31	...	...	17	15	12	2	0	4				
" 13	2	21	...	...	8	30	...	...	18	41	...	...	0	2	60	6	−24	5
	2	52	1	00	8	00	10	39	17	39	10	3	1	1				
" 14	3	24	1	30	7	30	10	38	16	38	12	1	1	3	60	0	−21	0
	3	53	0	30	8	00	9	06	16	36	10	3	1	3				
" 15	4	23	2	30	9	00	10	37	17	07	12	3	1	3	59	2	−16	5
	4	49	1	30	8	00	8	07	15	37	8	5	2	3				
" 16	5	16	4	0	9	45	11	11	16	56	11	7	2	5	58	3	−11	2
	5	40	2	0	9	00	8	44	15	44	8	9	3	7				
" 17	6	04	3	0	9	00	9	20	15	20	11	4	2	8	57	4	− 5	4
	6	27	2	30	11	29	8	26	17	25	9	3	4	6				
" 18	6	49	4	59	11	29	10	32	17	02	10	2	3	2	56	7	+ 0	4
	7	11	4	59	11	59	10	30	17	10	10	5	5	6				
" 19	7	33	5	59	...	...	10	48	...	...	10	5	...	...	56	0	+ 6	0
	7	54	7	29	1	29	11	56	19	18	10	0	4	0				
" 20	8	16	7	59	2	14	12	05	18	41	9	0	5	2	55	4	+11	3
	8	37	9	29	3	29	13	13	19	35	10	5	4	4				
" 21	8	59	8	59	2	44	13	22	18	28	9	4	5	8	54	9	+16	0
	9	22	9	29	2	59	12	30	18	22	13	0	3	0				
" 22	9	44	9	28	2	58	12	06	17	59	9	9	2	9	54	5	+20	1
	10	07	10	13	4	58	12	29	19	36	10	6	3	3				
" 23	10	31	9	28	3	58	11	21	18	14	8	5	3	6	54	3	+23	2
	10	55	11	28	5	43	12	57	19	36	11	0	2	9				
" 24	11	19	12	13	4	28	13	18	17	57	8	1	3	6	54	1	+25	4
	11	44	10	58	5	28	11	30	18	33	11	3	2	0				
" 25	...	...	11	58	4	58	12	14	17	39	8	3	2	6	54	0	+26	3
	0	10	...	...	6	58	...	...	19	14	...	...	3	1				
" 26	0	35	1	13	4	58	13	03	16	48	12	1	2	0	53	9	+26	0
	0	59	0	57	6	27	12	22	17	52	9	9	2	4				
" 27	1	24	...	...	4	57	...	...	15	58	12	3	1	9	54	0	+24	6
	1	49	0	27	7	27	11	03	18	03	9	8	3	0				
" 28	2	13	0	57	7	57	11	08	18	08	11	4	3	2	54	2	+22	1
	2	37	0	57	7	27	10	44	17	14	9	2	2	5				
" 29	3	00	1	12	7	57	10	35	17	20	11	4	3	6	54	5	+18	6
	3	24	1	42	6	27	10	42	15	27	0	1	2	0				
" 30	3	46	3	27	8	57	12	03	17	33	12	3	3	9	54	9	+14	3
	4	08	0	12	7	27	8	26	15	41	9	4	3	8				
July 1	4	30	4	42	9	57	12	34	17	49	11	6	2	6	55	5	+ 9	4
	4	52	4	57	8	57	12	27	16	27	9	3	5	0				
" 2	5	12	3	26	10	41	10	34	17	49	11	7	2	2	56	2	+ 4	0
	5	33	5	26	11	26	12	14	18	14	7	3	3	8				
" 3	5	55	4	11	10	26	10	38	16	53	9	1	4	8	57	0	− 1	7
	6	17	5	26	10	56	11	21	17	01	9	6	6	9				
" 4	6	39	6	41	11	11	12	24	16	54	10	2	4	4	57	9	− 7	5
	7	02	6	56	...	...	12	17	...	...	9	8	...	...				

Series III.—From April 20 to August 3, 1854.

Date. 1854.	Moon passes the meridian. Appar. time.		Apparent time of				Lunitidal interval.				Height of				Moon's parallax at noon.		Moon's declination at noon.	
			H. water.		L. water.		H. water.		L. water.		H. water.		L. water.					
	H.	M.	H.	M.	H.	M.	H.	M.	H.	M.	Ft.	Dec.	Ft.	Dec.	Min.	Dec.	Degree.	Dec.
July 5	7	26	6	56	0	26	11	54	17	47	10	1	6	5	58	8	—13	1
	7	51	7	56	1	26	12	30	18	24	10	6	4	3				
" 6	8	18	7	41	1	56	11	50	18	30	9	4	6	0	59	7	—18	3
	8	45	8	06	1	26	12	38	17	35	11	2	3	9				
" 7	9	14	7	55	2	55	11	10	18	37	9	2	5	6	60	5	—22	5
	9	43	...	...	3	10	...	...	18	25	...	...	3	0				
" 8	10	15	...	...	...	...	...	...	...	...	...	...	...	...	61	1	—25	3
	10	47	...	...	...	...	...	...	...	...	...	...	...	...				
" 9	11	20	10	40	4	40	11	53	18	25	8	5	2	4	61	4	—26	3
	11	54	11	40	4	25	12	20	17	38	12	5	1	2				
" 10	...	...	12	55	5	25	12	51	18	05	8	4	2	2	61	4	—25	4
	0	27	11	55	5	10	11	28	17	26	13	6	1	2				
" 11	1	00	12	25	6	25	11	25	17	58	10	1	1	5	61	0	—22	5
	1	31	12	55	6	10	11	24	17	10	13	3	1	2				
" 12	2	02	11	40	6	55	9	38	17	24	10	1	1	0	60	3	—18	3
	2	31	...	...	7	10	...	...	17	08	...	...	1	2				
" 13	3	00	1	10	7	40	10	39	17	09	13	6	1	0	59	5	—12	9
	3	26	1	25	7	55	10	25	16	55	10	0	2	4				
" 14	3	52	1	55	9	10	10	29	17	44	12	6	1	8	58	5	— 7	2
	4	15	2	24	8	54	10	32	17	02	9	2	2	5				
" 15	4	39	...	...	...	...	...	...	...	...	...	...	...	...	57	6	— 1	2
" 17	6	31	...	...	...	...	...	...	...	...	...	...	...	...			+10	2
" 18	6	53	...	...	...	...	...	...	...	...	...	...	...	...	55	3	+15	1
	7	16	7	24	1	24	12	31	18	53	9	2	4	8				
" 19	7	38	8	54	...	...	13	38	...	...	7	6	4	6	54	7	+19	3
	8	01	8	39	1	24	13	01	18	08	9	3	4	4				
" 20	8	24	9	54	2	39	13	53	19	01	8	7	5	6	54	4	+22	6
	...	...	...	...	1	54	...	...	17	53	...	...	4	5				
" 27	2	04	...	...	...	...	...	...	...	...	...	...	...	...			+15	3
" 28	2	26	...	...	...	...	...	...	...	...	...	...	...	...	55	2	+10	4
	2	48	1	54	7	54	11	28	17	28	8	6	2	0				
" 29	3	09	2	54	8	09	12	06	17	21	12	0	1	7	55	7	+ 5	2
	3	30	3	54	8	54	12	45	17	45	11	3	3	2				
" 30	3	51	2	09	8	24	10	39	16	54	11	1	3	2	56	3	— 0	4
	4	13	2	54	8	24	11	03	16	33	9	2	4	1				
" 31	4	34	...	...	...	...	...	...	...	...	10	6	4	0	56	9	— 6	1
	4	56	2	39	9	39	10	05	17	05	9	0	4	8				
Aug. 1	5	19	3	24	10	39	10	28	17	43	9	3	4	4	57	7	—11	6
	5	43	4	39	9	54	11	20	16	35	10	5	6	1				
" 2	6	07	4	54	11	09	11	11	17	26	10	5	4	5	58	5	—16	8
	6	32	5	54	...	...	11	47	...	...	9	5	...	...				
" 3	6	59	5	54	0	54	11	22	18	47	8	8	5	7	59	3	—21	3
	7	27	8	24	1	09	13	25	18	37	9	4	3	6				

Series IV.—From September 7 to October 22, 1854.

Sept. 7	0	23	...	...	...	...	...	...	...	...	...	...	...	...	59	3	— 5	7
	0	47	11	2	...	...	10	15	...	...	14	5	...	...				
" 8	1	12	11	2	5	32	9	50	16	45	14	0	—1	7	58	5	+ 0	5
	1	34	...	...	8	02	...	...	18	50	...	...	0	0				
" 9	2	01	1	32	8	02	11	56	18	26	13	5	1	0	57	7	+ 6	6
	2	24	1	32	7	32	11	31	17	31	14	0	—0	5				
" 10	2	47	1	03	7	33	10	39	17	09	13	0	—1	0	56	9	+12	3
	3	10	2	03	10	03	11	16	19	16	11	0	1	0				
" 11	3	33	2	35	8	33	11	23	17	23	11	0	0	0	56	1	+17	2
	3	57	...	...	9	03	...	...	17	30	10	5	1	5				
" 12	4	22	2	04	8	03	10	07	16	06	14	0	0	0	55	4	+21	2
	4	46	4	03	8	03	11	41	15	41	10	0	0	0				
" 13	5	11	3	04	7	04	10	18	14	18	10	0	0	0	54	8	+24	2
	5	35	3	34	8	04	10	23	14	53	9	0	3	0				
" 14	6	00	2	04	...	...	8	29	...	...	7	0	...	...	54	5	+26	1
	6	25	...	...	...	...	...	...	...	...	...	...	...	...				
" 15	6	52	...	...	...	...	...	...	...	...	...	...	...	...			+26	7
	7	17	...	...	...	...	...	...	...	...	...	...	...	...				

RECORD AND REDUCTION OF THE TIDES. 49

Series IV.—From September 7 to October 22, 1854.

Date. 1854.	Moon passes the meridian. Appar. time.		Apparent time of				Lunitidal interval.				Height of				Moon's parallax at noon.		Moon's declination at noon.	
			H. water.		L. water		H. water.		L. water		H. water.		L. water.					
	H.	M.	H.	M.	H.	M.	H.	M.	H.	M.	Ft.	Dec.	Ft.	Dec.	Min.	Dec.	Degree.	Dec.
Sept. 16	7	43	...	...	11	35	...	...	16	18	...	...	5	0	54	2	+26	1
	8	08	...	...	...	...	...	...	...	...	10	0	...	...				
" 17	8	34	7	03	1	33	11	58	17	53	9	0	5	0	54	3	+24	4
	8	58	8	06	0	06	11	32	15	58	10	4	5	0				
" 18	9	22	...	...	4	06	...	...	19	32	9	0	5	0	54	5	+21	5
	9	47	10	06	3	06	12	41	18	08	12	0	5	0				
" 19	10	09	8	36	4	36	10	49	19	14	10	0	5	0	54	9	+17	7
	10	33	11	06	6	06	12	57	20	19	12	6	5	0				
" 20	10	55	...	...	5	07	...	...	18	58	...	...	2	0	55	3	+13	1
	11	17	11	07	4	07	12	12	17	34	13	0	2	0				
" 21	11	39	11	07	5	07	11	50	18	12	11	5	4	0	55	8	+ 7	8
	...	...	11	07	3	07	11	28	15	50	14	0	0	0				
" 22	12	00	...	...	5	07	...	...	17	28	...	...	2	0	56	4	+ 2	2
	0	22	0	07	5	07	12	07	17	07	13	0	1	0				
" 23	0	45	1	08	6	08	12	46	17	46	12	0	-0	7	56	9	- 3	6
	1	07	0	08	6	38	11	23	17	53	13	0	0	0				
" 24	1	29	1	08	7	08	12	01	18	01	11	0	2	0	57	4	- 9	5
	1	52	0	08	7	08	10	39	17	39	13	0	1	0				
" 25	2	15	2	08	8	08	12	16	18	16	14	0	0	0	57	9	—14	9
	2	39	1	08	8	08	10	53	17	53	13	0	0	0				
" 26	3	04	1	09	7	09	10	30	16	30	13	0	0	0	58	3	—19	7
	3	32	2	09	8	09	11	03	17	03	13	0	0	0				
" 27	3	59	2	09	9	09	10	37	16	37	12	0	-1	0	58	6	—23	5
	4	27	3	09	8	09	11	10	16	10	13	0	0	0				
" 28	4	56	3	09	8	09	10	42	15	42	13	0	1	0	58	9	—25	0
	5	25	4	09	9	09	11	13	16	13	12	0	2	0				
" 29	5	57	2	10	10	10	8	45	16	45	10	0	2	0	59	2	—25	8
	6	28	4	10	9	10	10	13	15	13	11	5	3	0				
" 30	6	59	4	10	11	10	9	42	16	42	9	0	3	0	59	4	—25	8
	7	30	5	40	12	10	10	41	17	11	10	0	4	5				
Oct. 1	...	...	...	...	...	...	...	...	...	...	...	...	...	...	59	5	—23	3
" 3	10	20	...	...	...	...	...	...	...	...	...	...	...	...			—14	2
" 4	10	45	9	41	...	...	11	21	...	...	12	0	...	...	59	0	- 8	3
	11	10	11	11	4	41	12	26	18	21	14	0	0	0				
" 5	11	36	12	12	4	11	13	02	17	26	13	0	-1	2	58	6	- 2	0
	11	59	11	42	5	12	12	06	18	02	14	0	-2	0				
" 6	0	22	12	12	5	12	12	13	17	36	12	0	-1	0	57	9	+ 4	3
	...	...	11	12	7	12	10	50	19	13	14	0	-0	5				
" 7	0	46	...	...	6	42	...	...	18	20	...	...	1	0	57	3	+10	2
	1	09	0	12	5	42	11	26	16	56	14	0	0	0				
" 8	1	33	0	12	8	12	11	03	19	03	14	0	-1	0	56	6	+15	5
	1	57	0	43	7	13	11	10	17	40	13	0	0	0				
" 9	2	21	0	13	6	43	10	16	16	43	10	0	0	0	55	9	+20	0
	2	45	2	13	8	13	11	52	17	52	14	0	0	0				
" 10	3	10	1	13	6	13	10	28	15	28	13	0	0	0	55	3	+23	4
	3	35	2	13	8	13	11	03	17	03	14	0	0	5				
" 11	4	00	1	13	8	13	9	38	16	38	10	0	0	0	54	8	+25	7
	4	25	1	13	8	13	9	13	16	13	14	0	2	0				
" 12	4	51	2	13	8	13	10	48	15	48	14	5	2	0	54	5	+26	8
	5	17	3	44	9	14	10	53	16	23	14	0	4	0				
" 14	6	58	...	...	...	...	...	...	...	...	...	...	...	...	54	3	+25	3
" 15	7	23	...	...	11	14	...	...	16	16	10	0	3	0	54	4	+23	7
	7	47	8	14	11	14	12	51	15	51	10	0	5	0				
" 16	8	01	...	...	...	...	...	...	...	...	...	...	...	...	54	8	+19	2
	8	34	...	...	...	...	...	...	...	...	...	...	...	...				
" 17	8	57	10	15	3	15	13	41	19	14	10	0	3	0	55	2	+14	8
	9	19	10	15	3	15	13	18	18	41	11	0	3	0				
" 18	9	41	9	15	3	15	11	56	18	18	11	0	3	0	55	7	+ 9	7
	10	02	9	15	3	45	11	34	18	26	12	0	2	0				
" 19	10	24	9	15	4	15	11	13	18	34	12	0	0	0	56	4	+ 4	2
	10	46	...	...	3	45	...	...	17	43	11	0	1	0				
" 20	11	08	11	15	5	15	12	29	18	51	12	0	1	0	57	0	- 1	6
	11	30	12	15	4	15	13	07	17	29	11	0	0	0				
" 21	11	53	11	45	4	15	12	15	17	07	12	0	0	0	57	7	- 7	6
	...	...	11	15	5	15	11	22	17	45	12	0	0	0				
" 22	0	17	11	45	4	15	11	28	16	22	13	0	0	0	58	3	—13	4
	0	41	11	45	5	15	11	04	16	58	13	0	0	0				

7

The second form, or Table No. 2, for reduction of tides, is specially arranged to obtain the establishment and the half-monthly inequality in time and height. The first part is arranged in reference to the observed high waters; the second part, in reference to the low waters. That the inequality in time and height should also be made out from the low water, is specially important for stations where either the observations are of short extent, or else where difficulties tend to render the observations less accurate. The discussion of the low waters could not be omitted in our case. The headings to the columns of Table No. 2, explain the arrangement sufficiently. The results from the upper and lower transit of the moon are kept separate. (It need hardly be remarked that, in certain months, the sun's or moon's lower transit can be observed at Van Rensselaer Harbor.)

RECORD AND REDUCTION OF THE TIDES. 51

TABLE FOR THE REDUCTION OF TIDES.—No. 2.

Showing the Interval between the App. Time of the Moon's Superior Transit and the Time of High Water, and also the Heights of High Water, at Van Rensselaer Harbor, from Four Series of Observations made between October 10, 1853, and October 22, 1854.

0^h to 1^h.						1^h to 2^h.						2^h to 3^h.					
Moon's transit.	Lunitidal interval.		Height of H. water.		No. of observations and series.	Moon's transit.	Lunitidal interval.		Height of H. water.		No. of observations and series.	Moon's transit.	Lunitidal interval.		Height of H. water.		No. of observations and series.
App. time.	H. water.					App. time.	H. water.					App. time.	H. water.				
H. M.	H. M.		Ft. Dec.			H. M.	H. M.		Ft. Dec.			H. M.	H. M.		Ft. Dec.		
0 57	... ...		12 3			1 40	11 35		12 4			2 25	11 20		11 9		
0 10	10 57		11 6			1 16	12 00		11 8			2 16	11 00		11 6		I.
0 21	12 09		13 6		I.	1 07	11 23		13 0		I.	2 44	11 00		13 0		
0 53	11 17		9 5			1 55	11 20		12 1			2 58	10 34		... ...		
0 30	11 49		12 8			1 58	11 12		9 4								
												2 45	12 01		11 1		
0 01	12 16		11 9			1 01	11 15		9 0			2 29	11 32		13 1		
0 10	11 05		12 6			1 55	10 54		9 7			2 09	11 09		12 0		II.
0 58	11 48		13 6		II.	1 44	11 32		12 9		II.	2 53	11 10		10 6		
0 33	11 14		10 3			1 22	10 25		12 4			2 02	... ...		10 4		
0 20	10 52		12 7			1 15	12 06		8 4			2 52	11 00		9 5		
0 07	12 33		12 0			1 36	12 05		11 5								
0 52	13 03		12 5									2 45	10 33		12 6		
						1 09	11 24		13 0			2 33	11 16		12 3		
0 30	10 19		12 4			1 56	10 07		12 9			2 19	10 44		12 0		
0 39	11 54		10 7			1 30	10 34		11 3			2 21	10 39		10 3		III.
0 10	10 21		9 4		III.	1 20	10 34		10 9		III.	2 37	10 35		11 4		
0 10	13 03		12 1			1 16	10 15		10 4			2 02	9 38		10 1		
0 59	... ...		12 3			1 49	11 08		11 4			2 48	12 06		12 0		
						1 00	11 25		10 1								
0 22	12 46		12 0									2 01	11 31		14 0		
0 46	11 26		14 0		IV.	1 12	9 50		14 0			2 47	11 16		11 0		IV.
0 41	11 04		13 0			1 07	12 01		11 0		IV.	2 39	10 30		13 0		
						1 52	12 16		14 0			2 21	11 52		14 0		
						1 33	11 10		13 0								

MEANS.

| 0 29 | 11 40 | | ... 18 | | | 1 29 | 11 12 | | ... ... | 22 | | 2 32 | 11 04 | | ... ... | 20 |
| 0 31 | ... ... | | 12 1 | 20 | | 1 29 | ... ... | | 11 6 | 22 | | 2 29 | ... ... | | 11 8 | 20 |

Table for the Reduction of Tides.—No. 2.

Showing the Interval between the App. Time of the Moon's Superior Transit and the Time of High Water, and also the Heights of High Water, at Van Rensselaer Harbor, from Four Series of Observations made between October 10, 1853, and October 22, 1854.

\multicolumn{5}{c}{3ʰ to 4ʰ.}		\multicolumn{5}{c}{4ʰ to 5ʰ.}		\multicolumn{5}{c}{5ʰ to 6ʰ.}	
Moon's transit.		Lunitidal interval.		Height of H. water.		No. of observations and series.	Moon's transit.		Lunitidal interval.		Height of H. water.		No. of observations and series.	Moon's transit.		Lunitidal interval.		Height of H. water.		No. of observations and series.
App. time.		H. water.					App. time.		H. water.					App. time.		H. water.				
H.	M.	H.	M.	Ft.	Dec.		H.	M.	H.	M.	Ft.	Dec.		H.	M.	H.	M.	Ft.	Dec.	
3	12	12	03	11	1		4	00	11	16	9	9		5	42	12	04	9	7	
3	19	9	57	10	1		4	51	11	25	10	4		5	23	10	23	8	2	
3	34	10	40	11	5	I.	4	22	11	39	9	1	I.	5	14	10	29	10	1	I.
3	03	10	51	8	3		4	24	10	20	10	7		5	00	10	09	6	1	
3	46	11	01	10	4		4	03	10	36	9	1		5	52	12	46	6	8	
							4	32	11	14	11	5		5	17	11	14	11	6	
3	32	12	29	10	0		4	18	11	45	8	5		5	00	11	46	9	2	
3	14	...	...	11	7	II.	4	00	10	16	10	2	II.	5	43	12	33	10	5	
3	38	11	25	11	1		4	49	10	27	10	7		5	42	10	04	10	7	II.
3	41	10	53	10	7		4	23	11	25	9	2		5	09	11	09	10	5	
														5	56	10	53	8	8	
3	34	10	59	12	4		4	25	11	23	9	6		5	36	12	06	9	7	
3	38	10	56	10	4		4	41	10	08	10	6								
3	00	12	09	11	5		4	47	11	15	9	8		5	15	11	18	10	5	
3	59	10	19	11	0		4	23	9	07	8	5		5	40	10	09	9	6	
3	24	9	06	10	3	III.	4	08	12	34	11	6	III.	5	32	...	...	9	6	
3	24	12	03	12	3		4	52	10	34	11	7		5	16	8	44	8	9	III.
3	00	10	25	10	0		4	13	...	...	10	6		5	33	10	38	9	1	
3	52	10	32	9	2		4	56	10	28	9	3		5	43	11	11	10	5	
3	30	10	39	11	1															
3	33	...	...	10	5		4	22	11	41	10	0		5	11	10	23	9	0	IV.
3	32	10	37	12	0	IV.	4	27	10	42	13	0	IV.	5	25	8	45	10	0	
3	10	11	03	14	0		4	00	9	13	14	0								
							4	51	10	53	14	0								

MEANS.

| 3 | 30 | 10 | 57 | ... | ... | 19 | 4 | 27 | 10 | 52 | ... | ... | 21 | 5 | 27 | 10 | 53 | ... | ... | 19 |
| 3 | 29 | ... | ... | 10 | 9 | 21 | 4 | 27 | ... | ... | 10 | 5 | 22 | 5 | 27 | ... | ... | 9 | 5 | 20 |

The highest and lowest value of the interval balance nearly.

The criterion rejects no value of the interval, the two high and two low values balance nearly.

TABLE FOR THE REDUCTION OF TIDES.—No. 2.

Showing the Interval between the App. Time of the Moon's Superior Transit and the Time of High Water, and also the Heights of High Water, at Van Rensselaer Harbor, from Four Series of Observations made between October 10, 1853, and October 22, 1854.

6ʰ to 7ʰ.					7ʰ to 8ʰ.					8ʰ to 9ʰ.				
Moon's transit. App. time.	Lunitidal interval. H. water.	Height of H. water.		No. of observa- tions and series.	Moon's transit. App. time.	Lunitidal interval. H. water.	Height of H. water.		No. of observa- tions and series.	Moon's transit. App. time.	Lunitidal interval. H. water.	Height of H. water.		No. of observa- tions and series.
H. M.	H. M.	Ft.	Dec.		H. M.	H. M.	Ft.	Dec.		H. M.	H. M.	Ft.	Dec.	
6 28	13 45	8	0		7 26	12 32	6	7		8 22	11 51	7	9	
6 32	13 44	8	5		7 22	13 24	9	2		8 11	12 35	9	8	
6 21	10 55	7	8		7 14	12 02	8	1		8 59	12 47	10	8	
6 04	10 42	9	9		7 34	11 23	10	3		8 02	12 44	9	8	
6 48	11 55	8	5	I.	7 22	13 15	11	2	I.	8 48	13 13	9	5	I.
6 38	11 45	7	6		7 34	11 56	7	3		8 20	12 07	9	9	
6 04	10 29	8	8							8 04	13 33	11	1	
6 47	11 43	9	3							8 46	13 05	9	5	
					7 14	12 32	9	4		8 23	11 51	9	2	
					7 40	12 06	9	1	II.					
6 28	13 48	9	5	II.	7 42	12 41	8	5						
6 27	12 01	8	8		7 17	12 56	8	5		8 02	13 59	9	5	
										8 50	13 25	9	7	
6 45	11 16	8	1		7 41	12 05	7	2		8 41	12 05	8	9	II.
6 34	11 45	10	2		7 38	10 56	8	9		8 28	13 21	9	9	
6 16	...	9	4		7 24	13 10	8	7		8 42	13 57	9	2	
6 04	8 26	9	3	III.	7 00	13 02	9	7						
6 49	10 10	10	5		7 41	13 18	10	1	III.	8 35	13 12	9	8	
6 17	12 24	10	2		7 33	11 56	10	0		8 25	12 03	11	5	
6 53	12 31	9	2		7 02	11 54	10	1		8 10	12 24	8	6	
6 32	11 22	8	8		7 51	11 50	9	4		8 55	12 09	9	3	III.
					7 38	13 01	9	3		8 30	12 32	10	2	
6 28	9 42	9	0	IV.						8 16	13 13	10	5	
					7 43	...	10	0		8 59	12 30	13	0	
					7 23	12 51	10	0	IV.	8 45	11 10	9	2	
										8 34	11 32	10	4	IV.
										8 57	13 18	11	0	

MEANS.

| 6 30 | 11 35 | ... | ... | 18 | 7 28 | 12 26 | ... | ... | 20 | 8 32 | 12 42 | ... | ... | 21 |
| 6 29 | ... | 9 | 1 | 19 | 7 29 | ... | 9 | 1 | 21 | 8 32 | ... | 9 | 9 | 21 |

Peirce's criterion rejects the value 8ʰ 26ᵐ, new mean—

| 6 | 31 | 11 | 45 | ... | ... | 17 |

Table for the Reduction of Tides.—No. 2.

Showing the Interval between the App. Time of the Moon's Superior Transit and the Time of High Water, and also the Heights of High Water, at Van Rensselaer Harbor, from Four Series of Observations made between October 10, 1853, and October 22, 1854.

9^h to 10^h.						10^h to 11^h.						11^h to 12^h.					
Moon's transit.		Lunitidal interval.		Height of H. water.		Moon's transit.		Lunitidal interval.		Height of H. water.		Moon's transit.		Lunitidal interval.		Height of H. water.	
App. time.		H. water.			No. of observations and series.	App. time.		H. water.			No. of observations and series.	App. time.		H. water.			No. of observations and series.
H.	M.	H.	M.	Ft. Dec.		H.	M.	H.	M.	Ft. Dec.		H.	M.	H.	M.	Ft. Dec.	
9	12	12	47	9 1		10	02	12	12	9 8		11	26	12	50	12 2	
9	47	12	59	11 4		10	36	12	10	11 7		11	37	11	53	13 3	
9	31	12	30	11 2		10	13	12	03	11 5		11	50	12	21	10 8	I.
9	08	12	04	11 2		10	54	11	51	12 7		11	41	13	23	13 2	
9	58	13	13	12 5	I.	10	51	12	05	10 7	I.						
9	27	13	39	9 9		10	10	12	25	11 0		11	32	11	23	11 5	
9	18	12	10	9 1		10	54	12	11	12 5		11	41	12	06	11 3	II.
						10	16	12	42	10 6		11	19	12	06	10 9	
9	41	12	34	9 2		10	58	13	19	10 6		11	38	13	24	11 5	
9	45	12	32	9 7	II.	10	32	13	13	10 5		11	35	12	29	11 0	
9	18	13	32	10 6		10	45	12	02	10 4	II.	11	05	11	58	10 8	
9	39	12	15	9 6		10	31	12	53	11 4		11	51	12	34	12 5	III.
9	24	12	53	9 6		10	10	12	22	9 9		11	07	10	54	11 7	
9	10	11	24	11 3		10	43	12	21	10 5		11	19	11	39	11 3	
9	56	12	53	8 9	III.	10	21	13	27	12 9	III.	11	54	12	51	8 4	
9	38	12	41	11 4		10	31	12	57	11 0							
9	18	10	43	9 3		10	47	11	53	8 5		11	39	11	28	14 0	
9	44	12	29	10 6								11	10	13	02	13 0	
												11	59	12	13	12 0	IV.
9	22	12	44	12 0		10	09	12	57	12 6		11	08	13	07	11 0	
9	41	11	34	12 0	IV.	10	55	12	12	13 0		11	53	11	22	12 0	
						10	20	11	21	10 0	IV.						
						10	24	...	...	11 0							

MEANS.

| 9 | 31 | 12 | 30 | ... ... | 19 | 10 | 32 | 12 | 23 | ... ... | 20 | 11 | 32 | 12 | 17 | ... ... | 19 |
| 9 | 31 | ...| ...| 10 4 | 19 | 10 | 31 | ...| ...| 11 1 | 21 | 11 | 32 | ...| ...| 11 7 | 19 |

The value $10^h\ 43^m$ is rejected by Peirce's criterion, and there is no corresponding high value to balance it; the new mean becomes—

| 9 | 32 | 12 | 36 | ... | ... | 18 |

RECORD AND REDUCTION OF THE TIDES.

TABLE FOR THE REDUCTION OF TIDES.—No. 2.

Showing the Interval between the App. Time of the Moon's Inferior Transit and the Time of High Water, and also the Heights of High Water, at Van Rensselaer Harbor, from Four Series of Observations made between October 10, 1853, and October 22, 1854.

0ʰ to 1ʰ.					1ʰ to 2ʰ.					2ʰ to 3ʰ.				
Moon's transit. App. time.	Lunitidal interval. H. water.	Height of H. water.		No. of observations and series.	Moon's transit. App. time.	Lunitidal interval. H. water.	Height of H. water.		No. of observations and series.	Moon's transit. App. time.	Lunitidal interval. H. water.	Height of H. water.		No. of observations and series.
H. M.	H. M.	Ft.	Dec.		H. M.	H. M.	Ft.	Dec.		H. M.	H. M.	Ft.	Dec.	
0 35	11 10	10	7		1 18	11 12	11	5		2 02	11 13	11	5	
0 48	11 28	14	3		1 46	11 45	14	2		2 48	10 42	9	9	
0 44	11 01	10	9		1 31	11 14	13	7	I.	2 20	10 54	9	8	I.
0 21	11 50	12	8	I.	1 25	11 45	13	6		2 30	11 25	12	8	
0 06	11 58	12	5		1 44	11 19	...	...		2 34	11 29	...	...	
0 55	11 39	7	5											
					1 27	11 49	13	9		2 20	10 56	13	7	
0 31	11 31	13	3		1 21	11 10	11	8		2 06	10 10	13	1	
0 33	10 12	10	6		1 45	11 47	12	9	II.	2 51	12 25	12	6	II.
0 07	11 40	12	7		1 38	11 58	12	0		2 31	11 02	13	0	
0 57	11 35	13	0	II.	1 14	10 46	13	5		2 26	11 26	12	1	
0 06	12 15	11	2		1 59	11 57	13	0						
0 52	11 44	10	5							2 20	10 13	11	3	
0 29	12 11	14	5		1 33	10 30	12	0		2 01	10 48	13	9	
					1 00	10 04	13	6		2 44	11 34	8	5	
0 00	12 32	12	2		1 04	10 59	8	7		2 52	10 38	12	1	III.
0 45	12 18	11	6		1 54	10 54	8	5	III.	2 13	10 44	9	2	
0 02	11 32	13	1		1 49	...	...	12	2		2 31	10 39	13	6	
0 15	10 33	9	7	III.	1 31	11 24	13	3		2 26	11 28	8	6	
0 43	12 18	13	2											
0 35	12 22	9	9		1 36	11 56	13	5		2 24	10 39	13	0	
0 27	11 28	13	6		1 29	10 39	13	0	IV.	2 15	10 53	13	0	IV.
					1 09	11 03	14	0		2 45	10 28	13	0	
0 47	10 15	14	5		1 57	10 16	10	0						
0 00	12 07	13	0											
0 45	11 23	13	0	IV.										
0 22	10 50	14	0											
0 17	11 28	13	0											

MEANS.

| 0 29 | 11 34 | ... | ... | 25 | 1 31 | 11 13 | ... | ... | 21 | 2 27 | 10 59 | ... | ... | 20 |
| 0 29 | ... | 12 | 2 | 25 | 1 31 | ... | 12 | 3 | 21 | 2 27 | ... | 11 | 8 | 19 |

TABLE FOR THE REDUCTION OF TIDES.—No. 2.

Showing the Interval between the App. Time of the Moon's Inferior Transit and the Time of High Water, and also the Heights of High Water, at Van Rensselaer Harbor, from Four Series of Observations made between October 10, 1853, and October 22, 1854.

\multicolumn{6}{c	}{3^h to 4^h.}	\multicolumn{6}{c	}{4^h to 5^h.}	\multicolumn{6}{c}{5^h to 6^h.}
\multicolumn{2}{c	}{Moon's transit. App. time.}	\multicolumn{2}{c	}{Lunitidal interval. H. water.}	\multicolumn{2}{c	}{Height of H. water.}	No. of observations and series.	\multicolumn{2}{c	}{Moon's transit. App. time.}	\multicolumn{2}{c	}{Lunitidal interval. H. water.}	\multicolumn{2}{c	}{Height of H. water.}	No. of observations and series.	\multicolumn{2}{c	}{Moon's transit. App. time.}	\multicolumn{2}{c	}{Lunitidal interval. H. water.}	\multicolumn{2}{c	}{Height of H. water.}	No. of observations and series.
H.	M.	H.	M.	Ft.	Dec.		H.	M.	H.	M.	Ft.	Dec.		H.	M.	H.	M.	Ft.	Dec.	
3	36	9	39	10	5		4	25	10	51	10	3		5	16	10	15	8	7	
3	50	10	56	13	0		4	52	11	54	10	9		5	52	10	39	10	7	
3	09	10	35	9	1	I.	4	49	10	00	7	4	I.	5	37	12	36	7	1	I.
3	59	10	50	8	0		4	31	11	08	11	1		5	26	10	58	10	3	
3	52	10	37	13	7		4	09	10	53	11	6		5	39	11	52	7	5	
3	22	10	55	10	9		4	54	11	37	9	1								
3	08	11	38	12	9		4	38	11	08	8	6		5	21	11	55	8	9	II.
3	54	10	52	10	6		4	24	10	52	11	3		5	16	10	30	10	8	
3	37	11	24	10	6	II.	4	00	10	48	11	1	II.	5	32	11	47	9	6	
3	15	11	03	10	9		4	46	16	17	10	1		5	11	10	46	9	3	
3	18	11	04	12	3		4	12	12	10	12	4								
														5	10	10	39	12	5	
3	09	9	24	9	4		4	50	10	13	09	1		5	09	11	38	7	8	
3	59	10	34	10	3		4	09	10	40	12	8		5	54	10	08	8	1	
3	05	9	44	12	4		4	22	11	55	8	0	III.	5	40	9	20	11	4	III.
3	34	10	59	8	5		4	49	11	11	11	7		5	12	12	14	7	3	
3	53	10	37	12	3	III.	4	30	12	27	9	3		5	55	11	21	9	8	
3	00	10	42	9	1		4	34	10	05	9	0		5	19	11	20	10	5	
3	46	8	26	9	4															
3	25	10	29	12	6		4	46	10	18	10	0		5	35	8	29	7	0	IV.
3	09	12	45	11	3		4	56	11	13	12	0	IV.	5	57	10	13	11	5	
3	51	11	03	9	2		4	25	10	48	14	5								
3	10	11	23	11	0															
3	57	10	07	14	0															
3	05	11	03	13	0	IV.														
3	59	11	10	13	0															
3	35	9	38	10	0															

MEANS.

| 3 | 31 | 10 | 40 | ... | ... | 26 | 4 | 33 | 11 | 01 | ... | ... | 20 | 5 | 30 | 10 | 56 | ... | ... | 18 |
| 3 | 31 | ... | ... | 11 | 1 | 26 | 4 | 33 | ... | ... | 10 | 5 | 20 | 5 | 30 | ... | ... | 9 | 4 | 18 |

The two greatest deviations from the mean, viz., $8^h\ 26^m$ and $12^h\ 45^m$, nearly balance in the mean, hence no value was rejected.

The criterion rejects no value of the interval; the low value $8^h\ 29^m$ is so near the limit of rejection and not balanced in the mean that I prefer to reject it.

| 5 | 30 | 11 | 04 | ... | ... | 17 |

RECORD AND REDUCTION OF THE TIDES.

TABLE FOR THE REDUCTION OF TIDES.—No. 2.

Showing the Interval between the App. Time of the Moon's Inferior Transit and the Time of High Water, and also the Heights of High Water, at Van Rensselaer Harbor, from Four Series of Observations made between October 10, 1853, and October 22, 1854.

6^h to 7^h.						7^h to 8^h.						8^h to 9^h.					
Moon's transit.		Lunitidal interval.		Height of H. water.		Moon's transit.		Lunitidal interval.		Height of H. water.		Moon's transit.		Lunitidal interval.		Height of H. water.	
App. time.		H. water.				App. time.		H. water				App. time		H. water.			
H.	M.	H.	M.	Ft.	Dec.	No. of observations and series.	H.	M.	H.	M.	Ft.	Dec.	No. of observations and series.	H.	M.	H.	M.	Ft.	Dec.	No. of observations and series.
6	57	11	16	9	2		7	54	11	49	9	5		8	47	12	42	10	0	
6	07	10	54	7	1		7	46	12	30	8	1		8	35	...	...	9	3	
6	57	12	49	7	0		7	38	11	53	11	0		8	25	13	21	10	7	I.
6	48	11	58	10	4	I.	7	11	12	37	8	8	I.	8	44	12	58	10	9	
6	25	13	48	8	5		7	56	12	46	9	4		8	25	12	42	9	2	
6	14	13	39	10	6		7	42	12	25	11	8		8	50	12	38	11	2	
6	24	12	36	7	8		7	10	12	50	8	0								
							7	59	12	30	11	5		8	26	12	49	6	7	
6	06	11	10	10	7									8	10	14	06	10	7	II.
6	51	13	10	7	2	II.	7	38	11	53	8	2		8	53	13	57	8	8	
6	01	11	41	6	9		7	11	12	12	9	5	II.	8	12	11	42	10	2	
6	52	14	06	6	7		7	41	14	17	7	5								
														8	07	11	55	10	4	
6	07	11	12	11	7		7	13	12	18	10	0		8	59	11	18	10	7	
6	59	12	20	9	8		7	47	11	47	11	1		8	02	13	32	7	9	
6	38	12	54	8	6		7	22	13	40	9	2		8	48	13	31	8	1	
6	27	10	32	10	2	III.	7	11	10	48	10	5	III.	8	32	11	32	10	6	III.
6	39	12	17	9	8		7	54	12	05	9	9		8	07	11	55	10	4	
6	07	11	47	9	5		7	26	12	30	10	6		8	54	12	08	11	1	
6	59	13	25	9	4		7	16	13	38	7	6		8	37	12	22	9	4	
														8	18	12	38	11	2	
6	59	10	41	10	0	IV.	...	...	...	...	...	...	IV.	8	01	13	53	8	7	
6	58	...	...	10	0									8	08	11	58	9	0	
														8	58	...	...	9	0	IV.
														8	34	13	41	10	0	

MEANS.

| 6 | 33 | 12 | 13 | ... | ... | 19 | 7 | 33 | 12 | 28 | ... | ... | 18 | 8 | 29 | 12 | 44 | ... | ... | 21 |
| 6 | 34 | ... | ... | 9 | 1 | 20 | 7 | 33 | ... | ... | 9 | 6 | 18 | 8 | 30 | ... | ... | 9 | 7 | 23 |

There are two high and two low values, viz., $14^h\ 06^m$, $13^h\ 48^m$, and $10^h\ 32^m$, $10^h\ 41^m$, nearly balancing each other; there was, therefore, no rejection required.

The high and low values in the interval balance.

RECORD AND REDUCTION OF THE TIDES.

TABLE FOR THE REDUCTION OF TIDES.—No. 2.

Showing the Interval between the App. Time of the Moon's Inferior Transit and the Time of High Water, and also the Heights of High Water, at Van Rensselaer Harbor, from Four Series of Observations made between October 10, 1853, and October 22, 1854.

9^h to 10^h.					10^h to 11^h.					11^h to 12^h.				
Moon's transit. App. time. H. M.	Lunitidal interval. H. water. H. M.	Height of H. water. Ft. Dec.	No. of observations and series.		Moon's transit. App. time. H. M.	Lunitidal interval. H. water. H. M.	Height of H. water. Ft. Dec.	No. of observations and series.		Moon's transit. App. time. H. M.	Lunitidal interval. H. water. H. M.	Height of H. water. Ft. Dec.	No. of observations and series.	
9 37	12 37	10 9			10 24	12 20	11 3			11 01	12 30	12 0		
9 23	12 23	10 7			10 11	13 50	11 6			11 52	11 54	12 3		
9 09	12 22	11 8			10 34	12 41	11 0	I.		11 15	11 30	9 8		
9 52	11 54	11 5			10 24	13 17	13 8			11 59	12 01	13 0		I.
9 33	12 39	12 0	I.		10 51	11 49	8 6			11 20	11 36	13 7		
9 06	12 30	7 8								11 17	11 48	8 9		
9 49	13 17	9 6			10 06	13 39	6 6							
9 46	12 27	11 6			10 57	12 48	7 7			11 30	12 47	12 4		
					10 15	12 02	11 2	II.		11 46	11 29	8 4		
9 15	12 30	7 2			10 05	11 19	11 7			11 13	13 34	11 9		II.
9 12	12 05	10 3	II.		10 55	11 59	11 9			11 43	11 42	12 0		
9 11	12 43	10 3												
					10 32	11 30	12 2			11 16	12 46	12 0		
9 47	11 15	11 1			10 19	13 45	11 8			11 09	11 40	13 7		
9 33	11 31	9 4			10 43	11 35	10 0			11 28	13 17	9 7		
9 16	10 48	11 3	III.		10 38	10 23	12 9	III.		11 38	13 53	11 4		III.
9 59	10 49	9 9			10 07	11 21	6 5			11 44	12 14	8 3		
9 44	12 17	11 6			10 55	13 18	8 1			11 20	12 20	12 5		
9 22	12 06	9 9												
					10 45	12 26	14 0			11 17	11 50	11 5		
9 47	10 49	10 0			10 02	11 13	12 0	IV.		11 36	12 06	14 0		IV.
9 19	11 56	11 0	IV.		10 46	12 29	12 0			11 30	12 15	12 0		

MEANS.

| 9 30 | 12 03 | ... | ... | 19 | 10 29 | 12 18 | ... | ... | 19 | 11 28 | 12 16 | ... | ... | 19 |
| 9 30 | ... | ... | 10 4 | 19 | 10 29 | ... | ... | 10 9 | 19 | 11 28 | ... | ... | 11 5 | 19 |

There being three low and but one high value in the interval, it seemed preferable to adopt a mean resulting after the rejection of 10^h 48^m, viz:—

| 9 30 | 12 07 | ... | ... | 18 |

Table for the Reduction of Tides.—No. 2.

Showing the Interval between the App. Time of the Moon's Superior Transit and the Time of Low Water, and also the Heights of Low Water, at Van Rensselaer Harbor, from Four Series of Observations made between October 10, 1853, and October 22, 1854.

0ʰ to 1ʰ.					1ʰ to 2ʰ.					2ʰ to 3ʰ.				
Moon's transit. App. time.	Lunitidal interval. L. water.	Height of L. water.		No. of observations and series	Moon's transit. App. time.	Lunitidal interval. L. water.	Height of L. water.		No. of observations and series	Moon's transit. App. time.	Lunitidal interval. L. water.	Height of L. water.		No. of observations and series
H. M.	H. M.	Ft.	Dec.		H. M.	H. M.	Ft.	Dec.		H. M.	H. M.	Ft.	Dec.	
0 13	17 02	1	5		1 40	17 35	1	9		2 25	18 35	2	7	
0 57	17 33	2	0		1 16	18 00	0	0		2 16	17 45	0	3	
0 19	18 42	0	9	I.	1 07	17 38	3	1	I.	2 44	17 30	2	3	I.
0 21	18 54	3	1		1 55	18 20	1	2		2 58	17 04	...	...	
0 53	17 17	−2	1		1 58	17 12	−0	4						
0 30	18 49	1	6		1 20	15 58	0	6		2 45	18 01	1	3	
										2 29	17 47	2	3	
0 01	18 16	1	3		1 01	17 00	0	5		2 09	17 39	2	5	II.
0 10	17 05	4	1		1 55	17 21	−0	3		2 53	17 25	2	5	
0 58	18 18	2	6		1 44	18 02	4	1	II.	2 02	18 19	1	2	
0 33	17 44	0	1	II.	1 22	17 40	0	5		2 52	17 15	3	1	
0 29	19 37	2	7		1 15	17 21	3	0						
0 07	18 18	3	5		1 36	18 20	1	5		2 45	18 18	3	0	
0 52	19 03	3	3							2 33	17 46	0	5	
					1 09	18 24	2	1		2 19	17 59	2	5	
0 22	16 41	2	0		1 56	15 37	1	9		2 21	17 39	1	1	III.
0 30	16 49	1	3		1 30	16 49	0	8		2 37	17 20	3	6	
0 39	18 09	1	6	III.	1 29	17 34	1	5	III.	2 02	17 08	1	2	
0 10	18 21	2	0		1 16	17 15	0	4		2 48	17 21	1	7	
0 10	16 48	2	0		1 49	18 08	3	2						
0 59	15 58	1	9		1 00	17 10	1	2		2 01	17 31	−0	5	
										2 47	19 16	1	0	
0 22	17 46	−0	7	IV.	1 12	18 50	0	0		2 39	16 30	0	0	IV.
0 46	16 56	0	0		1 07	18 01	0	2	IV.	2 21	17 52	0	0	
					1 52	18 16	0	0						
					1 33	17 40	0	0						

MEANS.

| 0 30 | 17 49 | ... | ... | 21 | 1 29 | 17 34 | ... | ... | 23 | 2 31 | 17 43 | ... | ... | 21 |
| 0 30 | ... | ... | 1 | 7 | 21 | 1 29 | ... | ... | 1 | 3 | 23 | 2 30 | ... | ... | 1 | 7 | 20 |

The highest and lowest value of the Intervals balance in the mean, hence no value is rejected.

The low value 15ʰ 37ᵐ is rejected, hence new mean—

| 1 28 | 17 39 | ... | ... | 22 |

TABLE FOR THE REDUCTION OF TIDES.—No. 2.

Showing the Interval between the App. Time of the Moon's Superior Transit and the Time of Low Water, and also the Heights of Low Water, at Van Rensselaer Harbor, from Four Series of Observations made between October 10, 1853, and October 22, 1854.

3ʰ to 4ʰ.						4ʰ to 5ʰ.						5ʰ to 6ʰ.					
Moon's transit.		Lunitidal interval.		Height of L. water.		Moon's transit.		Lunitidal interval.		Height of L. water.		Moon's transit.		Lunitidal interval.		Height of L. water.	
App. time.		L. water.			No. of observations and series.	App. time.		L. water.			No. of observations and series.	App. time.		L. water.			No. of observations and series.
H.	M.	H.	M.	Ft.	Dec.	H.	M.	H.	M.	Ft.	Dec.	H.	M.	H.	M.	Ft.	Dec.
3	12	17	18	4	4	4	00	16	46	4	6	5	42	18	34	5	4
3	19	17	12	0	7	4	51	17	25	5	2	5	23	16	53	3	2
3	34	17	25	3	6	4	22	16	39	2	3	5	14	17	59	3	7
3	03	15	36	0	1	4	24	18	50	3	6	5	00	15	24	2	3
3	46	14	16	...	...	4	03	16	51	1	3	5	52	18	16	3	4
						4	32	16	59	4	4	5	17	17	44	4	2
3	32	17	29	2	0	4	16	16	30	3	4	5	00	16	01	3	8
3	14	18	02	0	3	4	00	17	16	3	0	5	43	18	33	5	0
3	38	16	10	3	9	4	49	17	27	3	0	5	09	16	54	6	1
3	44	16	55	2	6	4	23	16	25	3	1	5	56	16	53	5	8
												5	36	17	06	5	0
3	34	17	29	1	5	4	25	16	38	5	3						
3	38	15	41	2	3	4	41	15	38	2	0	5	40	15	39	4	2
3	09	18	09	2	5	4	47	17	45	3	7	5	32	...	...	4	5
3	59	17	48	4	0	4	23	15	37	2	3	5	16	15	44	3	7
3	24	17	36	1	3	4	08	17	49	2	6	5	33	16	53	4	8
3	24	17	33	3	9	4	52	17	49	2	2	5	43	17	26	4	5
3	00	16	55	2	4	4	13	...	...	4	0						
3	52	17	02	2	5	4	56	17	43	4	4	5	11	14	53	3	0
3	30	16	54	3	2							5	25	16	45	2	0
3	33	17	30	1	5	4	22	15	41	0	0						
3	32	16	37	−1	0	4	27	15	42	1	0						
3	10	17	03	0	5	4	00	16	13	2	0						
						4	51	16	23	4	0						

MEANS.

| 3 | 28 | 16 | 56 | ... | ... | 21 | 4 | 27 | 16 | 52 | ... | ... | 21 | 5 | 27 | 16 | 55 | ... | ... | 17 |
| 3 | 27 | ... | ... | 2 | 1 | 20 | 4 | 27 | ... | ... | 3 | 1 | 22 | 5 | 27 | ... | ... | 4 | 1 | 18 |

The low values 14ʰ 16ᵐ is rejected, hence new mean—

| 3 | 28 | 17 | 04 | ... | ... | 20 |

The high value 18ʰ 50ᵐ is in a measure balanced by two low values, 15ʰ 37ᵐ and 15ʰ 37ᵐ.

The low value 14ʰ 53ᵐ is rejected, hence new mean—

| 5 | 27 | 17 | 02 | ... | ... | 16 |

Table for the Reduction of Tides.—No. 2.

Showing the Interval between the App. Time of the Moon's Superior Transit and the Time of Low Water, and also the Heights of Low Water, at Van Rensselaer Harbor, from Four Series of Observations made between October 10, 1853, and October 22, 1854.

6^h to 7^h.						7^h to 8^h.						8^h to 9^h.					
Moon's transit.		Lunitidal interval.		Height of L. water.		Moon's transit.		Lunitidal interval.		Height of L. water.		Moon's transit.		Lunitidal interval.		Height of L. water.	
App. time.		L. water.				App. time.		L. water.				App. time.		L. water.			
H.	M.	H.	M.	Ft.	Dec.	H.	M.	H.	M.	Ft.	Dec.	H.	M.	H.	M.	Ft.	Dec.
6	28	16	45	4	7	7	26	17	47	4	8	8	22	18	21	4	2
6	32	17	59	5	2	7	22	18	24	4	8	8	11	18	50	4	5
6	21	17	55	4	0	7	14	18	02	4	4	8	50	19	02	3	4
6	01	18	32	3	7	7	34	17	38	4	2	8	02	18	29	5	5
6	48	17	55	3	7	7	22	18	45	5	2	8	48	18	43	4	2
6	38	18	00	2	9	7	34	18	10	3	2	8	20	19	22	2	7
6	01	17	29	2	9							8	04	18	03	5	3
6	47	18	28	2	4	7	14	19	47	5	6	8	46	18	50	1	8
						7	40	19	06	4	1	8	23	17	36	1	8
6	28	17	18	3	8	7	42	18	27	4	8						
6	27	20	01	6	0	7	17	19	11	5	5	8	02	19	13	3	3
												8	50	19	55	3	3
6	45	17	46	4	5	7	41	17	36	4	0	8	41	16	36	3	3
6	34	18	00	5	2	7	38	18	26	3	0	8	28	20	07	5	0
6	04	17	25	4	6	7	24	18	55	4	4	8	42	19	12	5	1
6	49	17	10	5	6	7	00	17	32	4	7						
6	17	16	54	4	4	7	44	18	48	4	0	8	35	18	27	4	1
6	53	...	...	4	6	7	33	18	41	5	2	8	25	19	39	2	7
6	32	18	37	3	6	7	02	18	24	4	3	8	10	17	24	3	4
						7	51	17	35	3	9	8	55	18	09	3	5
6	28	16	42	3	0	7	38	19	01	5	6	8	30	18	32	3	4
												8	16	18	28	5	8
						7	43	17	53	5	0	8	59	17	59	2	9
						7	23	15	51	5	0	8	45	18	26	3	0
												8	34	19	32	5	0
												8	01	19	14	3	0
												8	57	18	18	3	0

MEANS.

| 6 | 28 | 17 | 49 | ... | ... | 17 | 7 | 29 | 18 | 17 | ... | ... | 21 | 8 | 30 | 18 | 15 | ... | ... | 25 |
| 6 | 29 | ... | ... | 4 | 2 | 18 | 7 | 29 | ... | ... | 4 | 6 | 21 | 8 | 30 | ... | ... | 3 | 7 | 25 |

The high value $20^h 01^m$ is rejected, hence new mean—

| 6 | 28 | 17 | 42 | ... | ... | 16 |

The low value $15^h 51^m$ is rejected, hence new mean—

| 7 | 29 | 18 | 24 | ... | ... | 20 |

TABLE FOR THE REDUCTION OF TIDES.—No. 2.

Showing the Interval between the App. Time of the Moon's Superior Transit and the Time of Low Water, and also the Heights of Low Water, at Van Rensselaer Harbor, from Four Series of Observations made between October 10, 1853, and October 22, 1854.

9ʰ to 10ʰ.							10ʰ to 11ʰ.							11ʰ to 12ʰ.						
Moon's transit. App. time.		Lunitidal interval. L. water.		Height of L. water.		No. of observations and series.	Moon's transit. App. time.		Lunitidal interval. L. water.		Height of L. water.		No. of observations and series.	Moon's transit. App. time.		Lunitidal interval. L. water.		Height of L. water.		No. of observations and series.
H.	M.	H.	M.	Ft.	Dec.		H.	M.	H.	M.	Ft.	Dec.		H.	M.	H.	M.	Ft.	Dec.	
9	12	18	47	3	1		10	02	17	57	2	7		11	26	17	20	2	7	
9	47	17	59	2	9		10	36	17	40	1	6		11	37	18	53	3	5	I.
9	31	18	00	3	8		10	13	20	02	2	5		11	50	17	51	—1	2	
9	08	19	04	2	2	I.	10	54	18	36	2	0	I.	11	41	18	23	5	5	
9	58	18	43	0	7		10	51	17	50	0	1								
9	27	19	09	2	8		10	10	18	25	2	8		11	22	19	08	2	3	
9	18	18	10	1	9		10	54	19	11	2	4		11	41	17	21	—0	4	II.
														11	19	18	36	2	0	
9	41	19	34	1	5		10	58	18	19	1	0								
9	45	17	52	2	8	II.	10	32	18	13	1	8	II.	11	35	19	09	1	0	
9	39	18	45	3	8		10	45	18	02	1	0		11	35	16	59	0	9	
							10	31	18	23	2	2		11	05	17	58	2	7	
9	24	18	08	2	5									11	51	18	27	2	4	III.
9	10	18	09	1	8		10	10	19	22	1	6		11	07	18	54	0	6	
9	56	17	38	2	2		10	54	17	08	2	6		11	19	17	39	2	4	
9	38	19	10	4	0	III.	10	43	18	21	1	3		11	54	17	26	1	2	
9	18	17	13	1	8		10	21	18	42	2	0	III.							
9	44	18	14	3	6		10	10	16	51	1	4		11	39	17	28	2	0	
							10	31	17	57	3	6		11	10	18	02	—2	0	
9	22	19	14	5	0		10	47	17	38	1	2		11	59	19	13	—0	5	IV.
9	41	18	34	0	0	IV.								11	08	17	07	0	0	
							10	09	18	58	2	0		11	53	16	22	0	0	
							10	35	18	12	4	0	IV.							
							10	20	18	21	0	0								
							10	24	18	51	1	0								

MEANS.

| 9 | 32 | 18 | 28 | ... | ... | 18 | 10 | 32 | 18 | 19 | ... | ... | 22 | 11 | 32 | 18 | 01 | ... | ... | 19 |
| 9 | 32 | ... | ... | 2 | 6 | 18 | 10 | 32 | ... | ... | 1 | 9 | 22 | 11 | 32 | ... | ... | 1 | 3 | 19 |

TABLE FOR THE REDUCTION OF TIDES.—No. 2.

Showing the Interval between the App. Time of the Moon's Inferior Transit and the Time of Low Water, and also the Heights of Low Water, at Van Rensselaer Harbor, from Four Series of Observations made between October 10, 1853, and October 22, 1854.

\multicolumn{6}{c}{0ʰ to 1ʰ.}		\multicolumn{6}{c}{1ʰ to 2ʰ.}		\multicolumn{6}{c}{2ʰ to 3ʰ.}	
Moon's transit.		Lunitidal interval.		Height of L. water.		No. of observations and series.	Moon's transit.		Lunitidal interval.		Height of L. water.		No. of observations and series.	Moon's transit.		Lunitidal interval.		Height of L. water.		No. of observations and series.
App. time.		L. water.					App. time.		L. water.					App. time.		L. water.				
H.	M.	H.	M.	Ft.	Dec.		H.	M.	H.	M.	Ft.	Dec.		H.	M.	H.	M.	Ft.	Dec.	
0	35	17	25	1	8		1	18	17	27	1	6		2	02	15	43	5	7	
0	48	18	28	0	1		1	46	19	00	1	5		2	48	16	57	2	8	
0	44	18	16	1	3	I.	1	31	17	44	—0	1	I.	2	47	17	14	1	6	I.
0	21	18	05	—0	3		1	25	17	45	0	0		2	20	16	54	1	9	
0	06	16	58	3	4		1	44	17	19	...	...		2	30	18	10	—0	1	
0	55	17	08	2	7									2	34	17	58	...	...	
0	31	19	31	—1	8		1	27	18	19	—0	4		2	20	17	56	1	5	
0	33	18	13	2	4		1	21	17	25	3	2		2	06	17	40	2	2	
0	07	17	55	—0	9		1	45	19	02	1	3	II.	2	51	16	25	3	9	II.
0	57	18	05	1	1	II.	1	38	18	43	0	7		2	31	18	47	1	9	
0	06	18	15	2	3		1	14	18	27	1	3		2	26	18	56	1	6	
0	52	19	29	3	2		1	59	18	57	3	3								
0	29	17	56	—0	2									2	20	14	57	1	7	
0	00	17	02	2	2		1	33	16	00	1	5		2	01	17	33	1	7	
0	45	18	18	2	0		1	04	15	44	2	3		2	44	16	04	2	5	
0	15	16	03	0	5	III.	1	54	17	24	1	6	III.	2	52	16	38	1	6	III.
0	43	17	48	1	6		1	49	18	41	0	2		2	13	17	14	2	5	
0	35	17	52	2	4		1	24	18	03	3	0		2	31	17	09	1	0	
0	27	17	58	1	5		1	31	17	24	1	0		2	25	17	28	2	0	
0	47	16	45	—1	7		1	36	18	26	1	0		2	24	17	09	—1	0	
0	00	17	07	1	0		1	20	17	39	1	0	IV.	2	15	17	53	0	0	IV.
0	45	17	53	0	0	IV.	1	09	19	03	—1	0		2	45	15	28	0	0	
0	22	18	20	1	0		1	57	16	46	0	0								
0	17	16	58	0	0															

MEANS.

| 0 | 30 | 17 | 50 | ... | ... | 24 | 1 | 33 | 17 | 52 | ... | ... | 21 | 2 | 28 | 17 | 03 | ... | ... | 21 |
| 0 | 30 | ... | ... | 1 | 1 | 24 | 1 | 33 | ... | ... | 1 | 1 | 20 | 2 | 28 | ... | ... | 1 | 7 | 20 |

The two high and two low values of the intervals nearly balance in the mean.

The value 15ʰ 44ᵐ is rejected, hence new mean—

| 1 | 34 | 17 | 58 | ... | ... | 20 |

The lowest value 14ʰ 33ᵐ is rejected, hence—

| 2 | 28 | 17 | 15 | ... | ... | 20 |

TABLE FOR THE REDUCTION OF TIDES.—No. 2.

Showing the Interval between the App. Time of the Moon's Inferior Transit and the Time of Low Water, and also the Heights of Low Water, at Van Rensselaer Harbor, from Four Series of Observations made between October 10, 1853, and October 22, 1854.

	3ʰ to 4ʰ						4ʰ to 5ʰ						5ʰ to 6ʰ					
Moon's transit.		Lunitidal interval.		Height of L. water.		No. of observations and series.	Moon's transit.		Lunitidal interval.		Height of L. water.		No. of observations and series.	Moon's transit.		Lunitidal interval.		Height of L. water.		No. of observations and series.
App. time.		L. water.					App. time.		L. water.					App. time.		L. water.				
H.	M.	H.	M.	Ft.	Dec.		H.	M.	H.	M.	Ft.	Dec.		H.	M.	H.	M.	Ft.	Dec.	
3	36	17	40	3	4		4	25	16	06	3	9		5	16	15	15	4	5	
3	50	16	56	1	8		4	52	17	39	2	6		5	52	18	24	3	5	
3	09	17	05	2	6	I.	4	49	15	55	5	1	I.	5	37	16	51	6	6	I.
3	59	16	00	3	2		4	31	17	08	1	0		5	26	18	13	1	0	
3	32	17	37	1	6		4	09	16	22	4	5		5	39	17	51	3	1	
3	22	17	10	1	4		4	54	16	22	5	1								
														5	21	18	55	3	8	
3	08	17	53	2	1		4	38	16	53	2	2		5	16	16	00	5	5	II.
3	54	18	07	0	3		4	24	18	22	4	5	II.	5	32	17	02	4	1	
3	37	17	24	2	7	II.	4	00	16	48	1	0		5	11	17	01	4	2	
3	15	17	48	1	5		4	46	17	02	3	9								
3	18	16	34	2	7									5	10	17	39	1	8	
							4	50	16	13	6	5		5	09	17	08	4	8	
3	09	15	54	3	7		4	09	16	55	1	8		5	54	16	08	5	2	
3	59	16	34	4	3		4	22	17	40	3	6	III.	5	40	15	20	2	8	III.
3	05	16	59	1	8		4	49	16	56	2	5		5	12	18	14	3	8	
3	34	16	59	3	5		4	30	16	27	5	0		5	55	17	01	6	9	
3	53	17	07	1	3	III.	4	34	17	05	4	8		5	19	16	35	6	1	
3	00	15	27	2	0															
3	46	15	41	3	8		4	46	14	18	0	0		5	57	15	13	3	0	IV.
3	26	17	44	1	8		4	56	16	13	2	0	IV.							
3	09	17	45	3	2		4	25	15	48	2	0								
3	51	16	33	4	1															
3	10	17	23	0	0															
3	57	16	06	0	0															
3	06	17	03	0	0	IV.														
3	59	16	10	0	0															
3	35	16	38	0	0															

MEANS.

| 3 | 31 | 16 | 56 | ... | ... | 26 | 4 | 34 | 16 | 38 | ... | ... | 19 | 5 | 30 | 16 | 59 | ... | ... | 17 |
| 3 | 31 | ... | ... | 2 | 0 | 26 | 4 | 34 | ... | ... | 3 | 3 | 19 | 5 | 30 | ... | ... | 4 | 2 | 17 |

The value 14ʰ 18ᵐ is rejected, hence new mean—

| 4 | 34 | 16 | 45 | ... | ... | 18 |

The three highest and three lowest values of the intervals balance in the mean.

RECORD AND REDUCTION OF THE TIDES.

TABLE FOR THE REDUCTION OF TIDES.—No. 2.

Showing the Interval between the App. Time of the Moon's Inferior Transit and the Time of Low Water, and also the Heights of Low Water, at Van Rensselaer Harbor, from Four Series of Observations made between October 10, 1853, and October 22, 1854.

6ʰ to 7ʰ.						7ʰ to 8ʰ.						8ʰ to 9ʰ.					
Moon's transit.		Lunitidal interval.		Height of L. water.		No. of observations and series.	Moon's transit.		Lunitidal interval.		Height of L. water.		No. of observations and series.	Moon's transit.		Lunitidal interval.		Height of L. water.		No. of observations and series.
App. time.		L. water.						App. time.		L. water.						App. time.		L. water.				
H.	M.	H.	M.	Ft.	Dec.		H.	M.	H.	M.	Ft.	Dec.		H.	M.	H.	M.	Ft.	Dec.	
6	57	18	46	4	4		7	54	17	49	4	0		8	47	16	27	3	7	
6	07	16	39	5	2		7	46	17	00	5	5		8	35	19	11	5	1	
6	57	17	49	6	1		7	38	17	08	4	3		8	25	18	21	2	3	
6	48	17	58	4	0	I.	7	11	18	31	6	9		8	44	19	28	3	4	I.
6	25	17	48	3	4		7	56	18	46	4	3	I.	8	25	19	11	1	7	
6	14	18	54	3	5		7	00	18	07	3	6		8	50	19	08	2	9	
6	24	17	36	4	6		7	42	18	25	5	3								
							7	10	17	05	4	1		8	26	18	49	3	8	
6	00	17	40	4	5		7	59	18	45	4	8		8	10	20	36	5	7	
6	51	16	25	3	7									8	03	20	18	3	9	II.
6	10	17	36	4	3	II.	7	38	19	08	4	7		8	53	19	42	5	4	
6	01	16	41	5	7		7	09	21	07	4	7	II.	8	12	18	12	5	3	
6	52	17	08	5	5		7	11	18	42	5	3								
														8	07	19	10	4	0	
6	17	17	59	5	1		7	13	18	18	4	0		8	59	19	33	1	8	
6	07	17	57	3	0		7	14	17	35	6	0		8	02	17	02	5	2	
6	59	20	05	3	8		7	47	18	47	3	7		8	48	17	16	4	4	
6	38	17	24	5	6	III.	7	22	18	10	6	1	III.	8	32	18	47	2	5	III.
6	27	17	02	3	2		7	11	18	18	4	0		8	07	19	55	4	3	
6	39	17	47	6	5		7	54	19	35	4	4		8	54	17	53	4	0	
6	31	18	53	4	8		7	26	18	30	6	0		8	37	18	22	3	0	
6	07	18	47	5	7		7	16	18	08	4	4		8	18	18	37	5	6	
														8	01	17	53	4	5	
6	59	17	11	4	5	IV.	7	17	16	18	5	0	IV.	8	08	15	58	5	0	
6	58	16	16	3	0									8	58	18	08	5	0	IV.
														8	34	18	41	3	0	

MEANS.

| 6 | 31 | 17 | 45 | ... | ... | 22 | 7 | 28 | 18 | 18 | ... | ... | 21 | 8 | 29 | 18 | 41 | ... | ... | 24 |
| 6 | 31 | ... | ... | 4 | 6 | 22 | 7 | 28 | ... | ... | 4 | 8 | 21 | 8 | 29 | ... | ... | 4 | 0 | 24 |

The high value 20ʰ 05ᵐ is rejected, hence new mean— | The high value 21ʰ 07ᵐ is rejected, hence new mean— | The low value 15ʰ 58ᵐ is rejected, hence new mean—

| 6 | 30 | 17 | 39 | ... | ... | 21 | 7 | 29 | 18 | 10 | ... | ... | 20 | 8 | 30 | 18 | 48 | ... | ... | 23 |

Table for the Reduction of Tides.—No. 2.

Showing the Interval between the App. Time of the Moon's Inferior Transit and the Time of Low Water, and also the Heights of Low Water, at Van Rensselaer Harbor, from Four Series of Observations made between October 10, 1853, to October 22, 1855.

9ʰ to 10ʰ.						10ʰ to 11ʰ.						11ʰ to 12ʰ.					
Moon's transit.		Lunitidal interval.		Height of L. water.		Moon's transit.		Lunitidal interval.		Height of L. water.		Moon's transit.		Lunitidal interval.		Height of L. water.	
App. time.		L. water.				App. time.		L. water.				App. time.		L. water.			
H.	M.	H.	M.	Ft.	Dec.	H.	M.	H.	M.	Ft.	Dec.	H.	M.	H.	M.	Ft.	Dec.
9	37	18	37	2	7	10	11	19	05	3	9	11	01	18	00	1	9
9	23	18	53	4	9	10	34	18	26	2	1	11	52	19	09	3	6
9	09	19	07	2	8	10	24	17	47	1	2	11	15	18	15	0	7
9	52	18	54	2	6	10	31	18	04	2	3	11	59	17	01	0	4
9	33	19	08	2	6							11	20	18	06	1	7
9	06	19	00	0	3	10	06	18	54	1	3	11	17	17	48	2	4
9	49	18	16	3	5	10	57	18	03	2	8						
9	46	19	12	4	9	10	15	18	02	2	0	11	30	18	47	2	6
						10	05	18	34	3	4	11	46	17	59	0	7
9	15	21	00	3	7	10	55	18	30	1	3	11	13	18	49	3	0
9	12	18	35	2	8							11	43	18	42	-1	3
9	11	18	43	4	3	10	32	18	32	1	8						
						10	19	18	15	1	9						
9	47	18	45	1	7	10	43	18	50	2	0	11	16	19	16	0	1
9	33	20	01	2	0	10	38	18	23	2	1	11	09	18	40	1	0
9	16	19	03	3	3	10	07	19	36	2	9	11	28	19	05	3	4
9	59	18	19	2	0	10	55	18	33	2	0	11	38	18	23	1	8
9	22	19	36	3	3	10	15	18	25	2	4	11	44	19	14	3	1
												11	20	18	05	2	2
9	47	20	19	5	0	10	33	17	34	2	0	11	17	15	50	0	0
9	19	18	26	2	0	10	45	17	26	-1	2	11	36	17	36	-1	0
						10	02	17	43	1	0	11	30	17	45	0	0
						10	46	17	29	0	0						

Series: I., II., III., IV.

MEANS.

| 9 | 30 | 19 | 06 | ... | ... | 18 | 10 | 29 | 18 | 19 | ... | ... | 20 | 11 | 28 | 18 | 14 | ... | ... | 19 |
| 9 | 30 | ... | ... | 3 | 0 | 18 | 10 | 29 | ... | ... | 1 | 9 | 20 | 11 | 28 | ... | ... | 1 | 4 | 19 |

The highest value 21ʰ 00ᵐ is rejected, hence new mean—

| 9 | 30 | 19 | 00 | ... | ... | 17 |

The low value 15ʰ 50ᵐ is rejected, hence new mean—

| 11 | 28 | 18 | 21 | ... | ... | 18 |

The preceding tables (No. 2) contain the individual and mean values for interval and height, for high and low water, and the moon's upper and lower transit. The mean, in some cases, was improved by the application of Peirce's criterion for the rejection of doubtful observations; a few other rejections were made, as stated, in order to obtain a well-balanced mean; of 982 observations of the interval, but 17 were thus rejected.

Half-monthly Inequality.—For the comparison of the observed with the theoretical values, it is customary to use the forms of the equilibrium theory or of the wave theory,[1] certain modifications being necessary to produce an agreement between these theories with observation. According to the equilibrium theory the formula for the position of the pole of the tidal spheroid is:

$$\tan. 2\,\theta' = -\frac{h \sin. 2\,\phi}{h' + h \cos. 2\,\phi},$$

where h and h' are the elevations of the spheroid due to the sun and moon respectively, ϕ the angular distance of the moon from the sun and θ' the angular distance of the pole of the spheroid (or of high water) from the moon's place. In reality, however, the pole of this spheroid follows the moon at a certain distance, the mean value λ' of which is known as the "mean establishment" (also fundamental hour, corrected establishment), and which corresponds to a distance of the sun and moon of $\phi - \alpha$ instead of ϕ. This retroposition of the theoretical tide has been called the age of the tide. For the comparison of the observed and computed values for the half-monthly inequality *in time*, we have the formula:[2]

$$\tan. 2\,(\theta' - \lambda') = -\frac{h \sin. 2\,(\phi - \alpha)}{h' + h \cos. 2\,(\phi - \alpha)}.$$

This inequality goes through its period twice in each month. Proper values have to be found for the ratio $\frac{h}{h'}$ and the angle α.

The observations of 480 *high waters* furnish us with the following values, derived from the preceding tabulation on form No. 2:—

---

[1] An account of the Equilibrium, Laplace's and the Wave Theories, will be found in the Encyclopædia of Astronomy, forming a portion of the Encyclopædia Metropolitana, London, 1848; article "On Tides and Waves," by G. B. Airy, Esq., Astronomer Royal.

[2] Phil. Trans. Royal Society, 1834, Part I. On the Empirical Laws of the Tides in the Port of London, with some Reflections on the Theory; by the Rev. W. Whewell.

See also Phil. Trans. Royal Society, 1836, Part I. Researches on the Tides, fourth series: On the Empirical Laws of the Tides in the Port of Liverpool. By the Rev. W. Whewell.

From ☾'s upper transit.			From ☾'s lower transit.			From ☾'s upper and lower transit.		
Apparent solar time of moon's transit.	Lunitidal interval.	No. of observations.	Apparent solar time of moon's transit.	Lunitidal interval.	No. of observations.	Apparent solar time of moon's transit.	Lunitidal interval.	No. of observations.
0ʰ 29ᵐ	11ʰ 40ᵐ	18	0ʰ 29ᵐ	11ʰ 34ᵐ	25	0ʰ 29ᵐ	11ʰ 37ᵐ	43
1 29	11 12	22	1 31	11 13	21	1 30	11 12	43
2 32	11 04	20	2 27	10 59	20	2 30	11 01	40
3 30	10 57	19	3 31	10 40	26	3 30	10 47	45
4 27	10 52	21	4 33	11 01	20	4 30	10 56	41
5 27	10 53	19	5 30	11 04	17	5 29	10 58	36
6 31	11 45	17	6 33	12 13	19	6 32	11 59	36
7 28	12 26	20	7 33	12 28	18	7 30	12 27	38
8 32	12 42	24	8 29	12 44	21	8 30	12 43	45
9 32	12 36	18	9 30	12 07	18	9 31	12 22	36
10 32	12 23	20	10 29	12 18	19	10 30	12 21	39
11 32	12 17	19	11 28	12 16	19	11 30	12 16	38
Mean and sum	11 44	237	Mean and sum	11 43	243	Mean and sum	11 43.3	480

The mean establishment resulting from the observed times of 480 high waters at Van Rensselaer Harbor is therefore 11ʰ 43.3ᵐ, referred to the moon's transit immediately preceding and corresponding to a mean horizontal parallax of the moon and sun, and to the moon's and sun's declination of 16° nearly. The mean interval corresponds to the moon's transit of 0ʰ 21ᵐ nearly, indicating that the epoch would have come out 0ʰ 0ᵐ if transit E (see An Elementary Treatise on the Tides, by J. W. Lubbock, Esq., London, 1839) or that immediately preceding transit F had been used.

In like manner we obtain the following table from the observed times of 485 *low waters* at Van Rensselaer Harbor:—

From ☾'s upper transit.			From ☾'s lower transit.			From ☾'s upper and lower transit.		
Apparent solar time of moon's transit.	Lunitidal interval.	No. of observations.	Apparent solar time of moon's transit.	Lunitidal interval.	No. of observations.	Apparent solar time of moon's transit.	Lunitidal interval.	No. of observations.
0ʰ 30ᵐ	17ʰ 49ᵐ	21	0ʰ 30ᵐ	17ʰ 50ᵐ	24	0ʰ 30ᵐ	17ʰ 50ᵐ	45
1 28	17 39	22	1 34	17 58	20	1 31	17 48	42
2 31	17 43	21	2 28	17 15	20	2 30	17 29	41
3 28	17 04	20	3 31	16 56	26	3 30	16 59	46
4 27	16 52	21	4 34	16 45	18	4 30	16 49	39
5 27	17 02	16	5 30	16 59	17	5 29	17 00	33
6 28	17 42	16	6 30	17 39	21	6 30	17 40	37
7 29	18 24	20	7 29	18 10	20	7 29	18 17	40
8 29	18 15	25	8 29	18 39	23	8 30	18 31	48
9 32	18 28	18	9 30	19 00	17	9 31	18 43	35
10 32	18 19	22	10 29	18 19	20	10 30	18 19	42
11 32	18 01	19	11 28	18 21	18	11 30	18 11	37
Mean and sum	17 46	241	Mean and sum	17 50	244	Mean and sum	17 48.0	485

The mean establishment resulting from the observed times of 485 low waters is 17ʰ 48.0ᵐ, referred to the moon's transit immediately preceding low water, and the same to which the preceding high water has been referred; the difference between the two mean intervals is 6ʰ 04.7ᵐ.

To obtain a numerical expression for the half-monthly inequality in time, the value for α should be determined so as to furnish, in particular, good results for

RECORD AND REDUCTION OF THE TIDES.

$5^h\ 30^m$, $6^h\ 30^m$, $7^h\ 30^m$, where the curve is steepest; the value $\dfrac{h}{h'}$ is obtained from the greatest range of the inequality determined, for a first approximation, by a graphical process. I find from the observed high waters $a = 0^h\ 21^m$ or $5°\ 15'$, and from the low waters $a = 0^h\ 50^m$ or $12°\ 30'$. Range of inequality, from the high waters, $1^h\ 51^m$ or $27°\ 48'$, the *sin.* of which is 0.4649, and for the low waters, range $1^h\ 54^m$ or $28°\ 30'$, the *sin.* of which is 0.4771; hence the expression for the half-monthly inequality *in time* becomes

From the observed high waters $tan.\ 2(\theta' - 175°\ 49'.5) = - \dfrac{0.4649\ sin.\ 2(\varphi - 5°\ 15')}{1 + 0.4649\ cos.\ 2(\varphi - 5°\ 15')}$

" " " low waters $tan.\ 2(\theta' - 267°\ 00') = - \dfrac{0.4771\ sin.\ 2(\varphi - 12°\ 30')}{1 + 0.4771\ cos.\ 2(\varphi - 12°\ 30')}$

These expressions furnish us with the following comparison:—

HALF-MONTHLY INEQUALITY IN TIME.

	From high waters.				From low waters.		
Apparent solar time of moon's transit.	Observed.	Computed.	Difference.	Apparent solar time of moon's transit.	Observed.	Computed.	Difference.
$0^h\ 29^m$	$- 6^m$	$- 3^m$	$- 3^m$	$0^h\ 30^m$	$+ 2^m$	$+ 6^m$	$- 4^m$
1 30	−31	−22	− 9	1 31	0	−13	+13
2 30	−42	−39	− 3	2 30	−19	−31	+12
3 30	−56	−52	− 4	3 30	−49	−47	− 2
4 30	−47	−55	+ 8	4 30	−59	−56	− 3
5 29	−45	−38	− 7	5 29	−48	−52	+ 4
6 32	+16	+ 8	+ 8	6 30	− 8	−18	+10
7 30	+44	+46	− 2	7 29	+29	+33	− 4
8 30	+60	+56	+ 4	8 30	+43	+56	−13
9 31	+59	+48	− 9	9 31	+55	+54	+ 1
10 30	+38	+34	+ 4	10 30	+31	+42	−11
11 30	+33	+16	+17	11 30	+23	+25	− 2

Considering that the times of high and low water are only observed to the nearest half hour and for some time to the nearest hour, the agreement as shown above and by the diagrams, seems to be satisfactory.

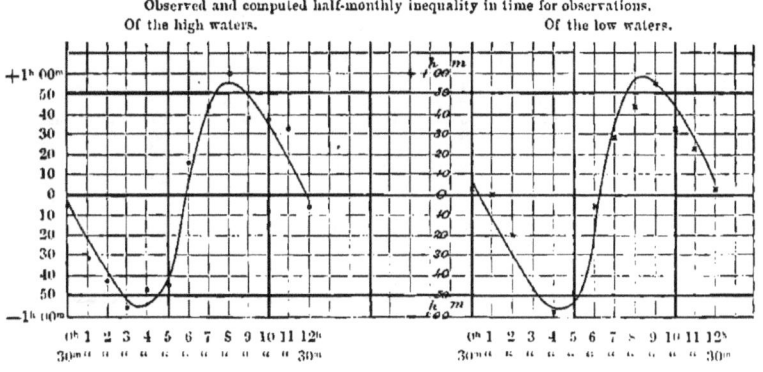

Observed and computed half-monthly inequality in time for observations.
Of the high waters.   Of the low waters.

In the above diagram, the observed values are indicated by dots; the computed values are represented by curves. From the times we have seen the mean value $\frac{h}{h'}$ (or $\frac{S''}{M''}$ of the wave theory and (A) of Lubbock's) = 0.471, and $a = 0^h\ 36^m$; hence, the age of the tide, or the time requisite for the moon to increase its right ascension by that amount, becomes $\frac{4}{5}$ days, or 18 hours.

Half-monthly Inequality in Height.—The theoretical expression for the half monthly inequality in height of high water is:

$$\eta = \sqrt{\{h'^2 + h^2 + 2\,h'h\cos.\ 2\,\phi\}}^1$$

where η expresses the height of the pole of the equilibrium spheroid above the mean level of the surface; for its application, and according to the wave theory, it must be changed to:

$$\eta = \sqrt{\{h'^2 + h^2 + 2h'h\cos.\ 2\,(\phi - a)\}}^2$$

The following table contains the results of the observations from the high and low waters, and the moon's superior and inferior transit:

From superior transits.			From inferior transits.			Means.	
Moon's transit.	Height of high water.	Number.	Moon's transit.	Height of high water.	Number.	Height of high water.	Number.
	Feet.			Feet.		Feet.	
0ʰ 31ᵐ	12.1	20	0ʰ 29ᵐ	12.2	25	12.1	45
1 29	11.6	22	1 31	12.3	21	11.9	43
2 29	11.8	20	2 27	11.8	19	11.8	39
3 29	10.9	21	3 31	11.1	26	11.0	47
4 27	10.5	22	4 33	10.5	20	10.5	42
5 27	9.5	20	5 30	9.4	18	9.5	38
6 29	9.1	19	6 34	9.1	20	9.1	39
7 29	9.1	21	7 33	9.6	18	9.3	39
8 32	9.9	24	8 30	9.7	23	9.8	47
9 31	10.4	19	9 30	10.4	19	10.4	38
10 31	11.1	21	10 29	10.9	19	11.0	40
11 32	11.7	19	11 28	11.5	19	11.6	38
		248			247	10.67	495

From superior transits.			From inferior transits.			Means.	
Moon's transit.	Height of low water.	Number.	Moon's transit.	Height of low water.	Number.	Height of low water.	Number.
	Feet.			Feet.		Feet.	
0ʰ 30ᵐ	1.7	21	0ʰ 30ᵐ	1.1	24	1.4	45
1 29	1.3	23	1 33	1.1	20	1.2	43
2 30	1.7	20	2 28	1.7	20	1.7	40
3 27	2.1	20	3 31	2.0	26	2.0	46
4 27	3.1	22	4 34	3.3	19	3.2	41
5 27	4.1	18	5 30	4.2	17	4.1	35
6 29	4.2	18	6 31	4.6	22	4.4	40
7 29	4.6	21	7 28	4.8	21	4.7	42
8 30	3.7	25	8 29	4.0	24	3.8	49
9 32	2.6	18	9 30	3.0	18	2.8	36
10 32	1.9	22	10 29	1.9	20	1.9	42
11 32	1.3	19	11 28	1.4	19	1.4	38
		247			250	2.72	497

[1] See Phil. Trans. Royal Soc., 1834 and 1836.

[2] Encyclopædia Metropolitana, Tides and Waves, Art. (535). The expression given by Mr. Lubbock is:

$$h = D + (E)\left\{(1 + \tfrac{\eta}{e})(A)\cos.(2\psi - 2\eta) + (1 + \tfrac{\eta'}{e})\cos.2\psi\right\};$$

for which see his treatise.

RECORD AND REDUCTION OF THE TIDES.

The values for h', h and α were found from the maxima and minima values of the inequality, viz., for the high water:

$$y = \sqrt{\{10.6^2 + 1.5^2 + 31.8 \cos. 2 (\phi - 15°)\}};$$

for the low waters:

$$y' = \sqrt{\{2.95^2 + 1.75^2 - 5.16 \cos. 2 (\phi - 15)\}};$$

These expressions may be changed to

$$y = 10.6 + 1.5 \cos. 2 (\phi-15°), \text{ and } y' = 2.7 - 1.7 \cos. 2 (\phi-15°);$$

they leave the following differences between the computed and observed values:—

Moon's transit.	Height of high water.			Height of low water.		
	Computed.	Observed.	Difference.	Computed.	Observed.	Difference.
0ʰ 30ᵐ	12.1	12.1	0.0	1.1	1.4	+0.3
1 30	12.0	11.9	—0.1	1.1	1.2	+0.1
2 30	11.7	11.8	+0.1	1.5	1.7	+0.2
3 30	11.0	11.0	0.0	2.3	2.0	—0.3
4 30	10.2	10.5	+0.3	3.1	3.2	+0.1
5 30	9.5	9.5	0.0	3.9	4.1	+0.2
6 30	9.1	9.1	0.0	4.3	4.4	+0.1
7 30	9.2	9.3	+0.1	4.3	4.7	+0.4
8 30	9.5	9.8	+0.3	3.9	3.8	—0.1
9 30	10.2	10.4	+0.2	3.1	2.8	—0.3
10 30	11.0	11.0	0.0	2.3	1.9	—0.4
11 30	11.7	11.6	—0.1	1.5	1.4	—0.1

The differences may be considered within the uncertainty of the observations. The annexed diagram shows the comparison given above:—

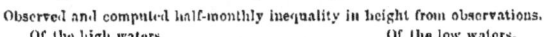
Observed and computed half-monthly inequality in height from observations.
Of the high waters.   Of the low waters.

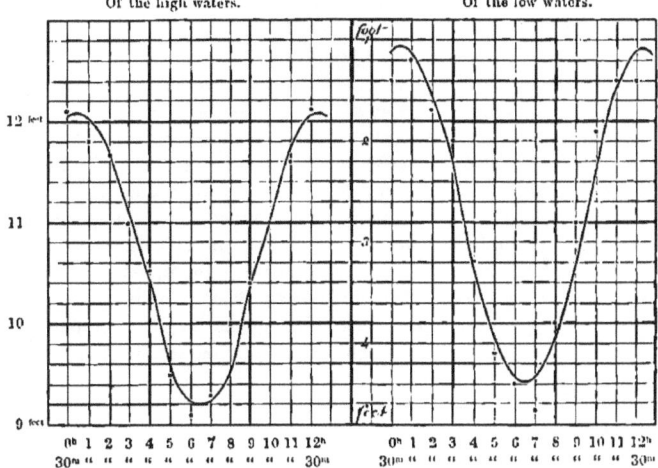

From the inequality in height $\dfrac{h}{h'}$ or $\dfrac{S'''}{M'''}$ (notation of the wave theory) $= 0.367$

whereas from the inequality in times $\dfrac{h}{h'}$ or $\dfrac{S''}{M''}$. $= 0.471$

The ratio[1] of the solar to the lunar tide is deduced with more exactness from the inequality in times, and the above value is certainly greater than the average value deduced at more southern stations. One of the reasons why this ratio is not constant, and which probably applies here, is given in (538, β) (Tides and Waves), viz.: If tides are communicated by different channels to the same port, the proportion of the solar and lunar waves will depend on the length of those channels. This explanation would require a polar tide to enter through Kennedy Channel, to combine with the principal tide which passes up Baffin's Bay, and enters by Smith's Straits. According to the equilibrium theory, there should be no tide at the pole, and but a small tide in latitude $78\frac{1}{2}°$; but it is the tide wave propagated from the Atlantic, which is felt in this part of the polar regions. With regard to a, its value as found by the heights is more accurate than that found by the times; the latter gave $a = 9°$, the former $15°$ (the same from high and low waters). Adopting $15°$, the retard or age of the tide becomes $1\frac{1}{4}$ day, by which interval the spring and neap tides follow the syzygies and quadratures, respectively. The time-value of a is here smaller than the height-value, which is more in accordance with theory than the opposite, as observed at a number of places on the coast of England (543 and 546, Tides and Waves). Compared with other values of a, the Van Rensselaer value appears somewhat smaller than an average at more southern stations.

We have further, mean rise and fall of tides at Van Rensselaer Bay 7.9 feet, range of spring tides 11.1 feet, and range of neap tides 4.7 feet. These numbers are averages from the discussions of $9\frac{1}{2}$ lunations, and obtain without regard to the diurnal inequality, which will be investigated further on.

Effect of the Changes in the Moon's Declination and Parallax on the half-monthly Inequality, in Time.—In reference to the investigation of the half-monthly inequality, it is comparatively of little consequence which transit of the moon is taken for comparison; it is otherwise in the investigation of the effect of a change in the moon's declination and parallax, as well as for a similar effect due to the sun, which latter, however, cannot become a subject of investigation for the tidal series in hand, on account of its short extent; for the same reason, the variation in the inequality, in height, will have to be passed over. To ascertain the effect due to the moon's declination and parallax, an anterior value, corresponding to a certain age of the tide, is to be taken in the comparison; the preceding investigation gave for the retard $1\frac{1}{4}$ day, each lunitidal interval, minus its corresponding mean value for the respective hour of the moon's transit, was therefore tabulated in respect to the moon's declination and parallax (separately for each), corresponding to one day anterior to the time of high or low water, thus referring the results to transit E. The present investigation can only furnish an approximation to the true results; the

---

[1] For comparison of different values for this ratio, the following have been selected: $\frac{S''}{M'}$ for London, 0.379; for Plymouth, 0.407; from the discussions of the Superintendent of the U. S. Coast Survey, for Key West, 0.325; San Diego, 0.39; and San Francisco, 0.342. (*Annual Reports* of 1853 and '54.) $\frac{S'''}{M'''}$ for Dundee, 0.277; for Brest, 0.346; for Plymouth, 0.294.

observations, while they give reliable value for the half-monthly inequality, cannot be expected to give more than an approximation to its variations. For any one station, and any one inequality or correction to it, special examinations require to be made to ascertain that transit of the moon, best suited for the purpose; this has hardly been done for any standard station, and it suffices to state here that, by referring to an anterior transit, the whole half-monthly inequality is moved backward through nearly twenty-four minutes for every transit preceding. Upon the inequality itself, the effect is but of a differential character. Thus to refer our table to transit E, deduct 24^m from each value.

To concentrate as many values as possible to a mean, the changes of declination and parallax were grouped for three values. The separate parcels for declination are for declination 0 to 13°, 13° to 21°, and 21° to 27°.5, irrespective of sign. The parallax groups are: 54′ to 56′, 56′ to 58′, and 58′ to 61′.4.

The differences of interval for the high and low waters were made out separately, and, in general, agreed tolerably well. I obtained the following results:—

TABLE SHOWING THE CORRECTION (IN MINUTES) TO THE MEAN HOURLY INTERVAL, FOR A CHANGE IN THE MOON'S DECLINATION AND PARALLAX.

Moon's transit.	Correction to interval for moon's declination.			Correction to interval for moon's parallax.		
	0 to 13°.	13° to 21°.	21° to 27°.5.	54′ to 56′.	56′ to 58′.	58′ to 61′.4.
0^h 30^m	− 2^m	− 7^m	+ 5^m	−12^m	+14^m	− 7^m
1 30	+15	− 5	−18	−17	+17	+ 5
2 30	+21	− 7	−12	−11	+12	− 2
3 30	+23	−11	− 9	− 1	+20	−10
4 30	+ 9	0	−13	− 1	+ 1	− 3
5 30	+14	+ 4	−15	+16	+19	−36
6 30	− 9	+18	− 3	+ 7	− 3	− 2
7 30	+ 1	−13	+ 9	+ 5	− 3	− 6
8 30	− 2	+ 9	+15	+26	−10	+ 8
9 30	− 6	+ 1	+22	+ 9	−11	− 5
10 30	+ 1	− 6	+10	+10	− 9	− 7
11 30	− 3	− 2	+ 3	− 4	+ 4	+ 4
Mean	+ 5^m	− 1^m	− 1^m	+ 2^m	+ 4^m	− 5^m
No. of observ.	373	262	348	387	214	333

Mean declination 16°.0.   Mean parallax 57′.0.

The above table of declination corrections exhibits systematic values for the periodical part of the lunar effect, or for the term $D \sin. 2 (\phi - \gamma)$. Between 0° and 13° of declination, the correction is positive for transits between 1^h and 7^h, for other hours negative; for declinations between 13° and 21° it is positive, between the hours of 4 and 10; for remaining hours it is negative, and for declinations 21° to 27°.5, the correction is positive, for hours 7 to 1, and negative for remaining hours of transit. The quantity D is accordingly about 14 minutes, and γ equals 15°, 60°, and 105° respectively.

The variation in the inequality due to the changes of the moon's declination appears large when compared with its value at other places, but is in conformity with the large value of the half-monthly inequality itself.

The periodical part of the parallax correction is of the same form as given above.

The empirical values for the groups of small and middle values of parallax appear systematic; the values in the last column for large parallax are less regular. The maximum correction on the average is somewhat greater than one-fourth of an hour.

The corrections to the mean establishment for changes of the sun's declination and parallax may be taken as one-third of the corresponding lunar values, and in the present case will probably not exceed five minutes of time.

The means of each column, containing the non-periodical part, are small, and appear rather irregular; they are variable with the transit or the moon's age adopted in the discussion.[1]

Diurnal Inequality.—We now proceed to the examination of a prominent feature in the Rensselaer Harbor tides, namely, the diurnal inequality. This inequality is well marked in the diagrams, Plates I, II, and III. Although the existence of this inequality, in height and times, has long been known to practical men, it was not until about twenty-five years ago that its laws were understood and reduced to computation by Mr. Whewell.[2] The subject has since been taken up by the present superintendent of the U. S. Coast Survey, Prof. Bache;[3] his researches commenced about nine years ago, and resulted in a further extension of the method of discussion as well as in the recognition of the geographical limits of the phenomena on our own coast; further, the discussion of single day tides, produced by this inequality in extreme cases, and here complicated by an extremely small rise and fall of the tides, was now successfully accomplished. According to the equilibrium theory, the diurnal tide ought to be very small in latitude 79°; but viewing the Rensselaer Harbor tide as a wave, produced principally in the Atlantic, and propagated through Davis's and Smith's Straits, the existence of the diurnal inequality in so high a northern latitude cannot surprise us. The following notes were extracted from Captain McClintock's narrative of the voyage of the "Fox,"

---

[1] On this point the reader may consult Whewell's 9th series of tidal researches: "Laws of the Tides from a Short Series of Observations," Phil. Trans. 1838; also Airy, "Tides and Waves," articles 552 and following.

[2] Researches on the Tides, sixth series. On the Results of an Extensive System of Tide Observations made on the Coasts of Europe and America in June, 1835. By the Rev. W. Whewell. Phil. Trans. Roy. Soc. 1836.

Researches on the Tides, seventh series. On the Diurnal Inequality of the Height of the Tide, especially at Plymouth and Singapore. By the same author. Phil. Trans. 1837.

Researches on the Tides, eighth series. On the Progress of the Diurnal Inequality Wave along the Coasts of Europe. By the same author. Phil. Trans. Roy. Soc. 1837.

[3] Note on a Discussion of Tidal Observations at Cat Island in the Gulf of Mexico, by Prof. A. D. Bache. Coast Survey Report for 1851, App. No. 7; Additional Notes thereto, Coast Survey Report for 1852, App. No. 22.

On the Tides at Key West and of the Western Coast of the United States. Coast Survey Report for 1853, App. Nos. 27 and 28. By Prof. A. D. Bache.

Comparison of the Diurnal Inequality of the Tides at San Diego, San Francisco, and Astoria, on the Pacific Coast of the United States. Coast Survey Report for 1854, App. No. 26. By Prof. A. D. Bache.

Approximate Co-Tidal Lines of Diurnal and Semi-Diurnal Tides of the Coast of the United States on the Gulf of Mexico. Coast Survey Report for 1856, App. No. 35. By Prof. A. D. Bache.

For the theoretical investigation of the diurnal tide, see also Airy's Tides and Waves, articles 46 and following; and articles 562 and following.

in 1857, '58, '59. Referring to Bellot Strait: "As in Greenland, the night tides are much higher than the day tides." Speaking of the ice motion, and remarking that the tides are the chief cause of it, he says: "Now we know that the night tides in Greenland greatly exceed the day tides." Also, when near Buchan Island, north of Upernavik, and in the vicinity of Cape Shackleton: "We had grounded during the day tide, and were floated off by the night tide, which on this coast occasions a much greater rise and fall." By the labors of Dr. Kane we now know that the diurnal inequality extends as high up as 79° of latitude on the northwestern coast of Upper Greenland. In a report of Mr. Sonntag's to Dr. Kane, dated Godhavn, Sept. 12, 1855, he says: The mean height of spring tides is 12.8 feet, and at the time of new and full moon high water is at $12^h\ 0^m$; the highest spring tide is three days after full moon, and the night tide is at this time fully three feet higher than the day tide. At Northumberland Island, Sept. 10, 1854, at (after) the time of full moon high water was at 11^h P. M., and the night tide rose three feet more than the day tide. These statements, crude as they necessarily are, show that the attention of the party was fully directed to the phenomenon.

A cursory examination of the Plates (I, II, and III) shows that the diurnal inequality extends without exception over the whole series of observation, that it is well marked in the difference of the height of high water, but very little or irregularly in the height of low water; that sometimes the day tide, at other times the night tide is the higher of the two occurring in a lunar day; further, that it vanishes a day or two after the moon's crossing the equator, and that it amounts in maximo to about three feet some time after the moon attains her greatest declination. There is but one instance where the inequality approximates to the production of a single day tide. See curve for Nov. 23, 1853.

We may now enter somewhat more fully into the discussion of this inequality, which is produced by the interference of two independent waves, the diurnal and the semi-diurnal, the former depending for its size chiefly on the moon's declination. For a complete study of these compound waves, they require to be examined in their separate parts, and it would therefore be our first object to effect their separation into the diurnal and the semi-diurnal; a process which, when graphically performed, is neither too laborious nor lacking in accuracy; it is nevertheless a process of some nicety, and requires observations of standard excellence. Upon trial, I found the less rigorous method employed by Mr. Whewell in his discussion of the Plymouth and Singapore tides, was better suited to the general mass of the observations at Van Rensselaer, and that the above described process of separation had better be reserved to that portion of our observations which are apparently of the best character.

The observed heights of high and low water were laid down graphically, and a line was drawn by the eye, cutting off the zigzags of the successive high waters, leaving equal portions above and below the intermediate curve. These differences from the mean height were then set off from another axis, and those belonging to the high water next following the moon's superior transit were marked by a curve of dashes; those following the moon's inferior transit were marked by a curve of dots. These curves, without exception, were found to have alternately, as the moon has north or south declination, positive and negative ordinates, in perfect accordance with the equilibrium theory, according to which the tide (high water) which belongs to a south transit of the moon should be the greater of the two of the same day, the moon's declination being north, or should be the smaller of the two, the moon having south declination; when the moon crosses the equator (or, according to experience, some time after it), the inequality vanishes; the time by which the full effect is produced is, as in other cases of the application of this theory, later than theoretically indicated. On Plate III are given specimens of the diurnal inequality curve, constructed as explained above and on the same scale as the other diagrams on these plates. By means of the diagrams, the epoch when the inequality vanishes has been made out as follows:—

TABLE SHOWING THE OBSERVED TIMES WHEN THE DIURNAL INEQUALITY VANISHES, TOGETHER WITH THE TIME WHEN THE MOON CROSSES THE EQUATOR, AND THE DIFFERENCE OF THESE TIMES, OR THE NUMBER OF DAYS BY WHICH THE CAUSE PRECEDES THE EFFECT. THIS DIFFERENCE IS ALSO CALLED THE EPOCH.

Year.	Inequality disappears.	Moon's declination equal 0.	Difference, or epoch.	Year.	Inequality disappears.	Moon's declination equal 0.	Difference, or epoch.
1853	Oct. not observed	15^d 7^h	...	1854	April 23^d 8^h	24^d 11^h	3^d 21^h
"	Oct. 30^d 21^h	29 18	1^d 3^h	"	May 9 24	9 0	1 0
"	Nov. 12 10	11 13	1 7	"	" 23 14	21 17	1 21
"	" 13 6			"	June 7 9	5 10	1 23
"	" 27 22	26 4	1 18	"	" 19 9	17 22	1 11
"	Dec. 9 9	8 19	0 14	"	July 5 4	2 17	2 11
"	" 25 12	23 13	1 23	"	" 31 22	29 22	2 0
1854	Jan. not observed.	5 2	...	"	Sept. 9 3	7 22	1 5
"	" " "	19 18	...	"	Remaining observations of Series IV not sufficiently reliable.	22 9	...
"	Feb. 5 4	1 10	1 18			Mean.	1^d 15^h
"	" 14 18	15 23	0 19				
"	Mar. 3 12 } 4 12 }	28 19	1 17				
"	" 18 0	15 5	0 19				
"	Mar. obs'n incompl.	28 4	...				

The results for the epoch are very regular, and with the exception of part of the last series, which is of inferior accuracy, no observation has been omitted. The inequality vanishes at the distance of 1.62 days' motion of the moon from her nodes.

The magnitude of the diurnal inequality, and its variation depending on twice the moon's declination, was made out by dividing the inequality curves in six parts between the times of disappearance, and by tabulating the ordinates as well as the corresponding declination of the moon, the following results were obtained from 12 complete cycles, omitting no value, viz:—

AMOUNT OF DIURNAL INEQUALITY IN THE HEIGHT OF HIGH WATER.

Ordinate.	(In feet.)												Mean dh.	Mean declination.
0	0	0	0	0	0	0	0	0	0	0	0	0	0.0	0°
1	1.1	2.1	0.8	2.7	1.0	0.3	0.2	1.4	2.7	1.5	1.5	1.0	1.4	12
2	0.5	2.9	2.3	4.0	4.2	2.5	1.0	1.5	2.2	2.0	2.1	3.0	2.3	21
3	1.5	2.2	3.0	4.6	4.0	2.8	1.1	1.6	2.2	3.1	2.0	2.0	2.5	25
4	1.6	2.3	3.0	4.6	0.2	3.0	1.5	1.4	1.8	1.9	3.3	2.7	2.3	22
5	0.8	1.1	1.1	3.0	2.4	1.2	0.7	3.5	2.2	1.0	2.0	2.0	1.7	13
6	0	0	0	0	0	0	0	0	0	0	0	0	0.0	0

RECORD AND REDUCTION OF THE TIDES.   77

The mean declination corresponds to an epoch 1.6 days anterior, which remark applies also to the formula $dh = C \sin. 2 \delta'$, representing the diurnal inequality dh in two successive high or low waters, δ' being the moon's declination. For the value of C we obtain 3.3, which gives us the following comparison:—

DIURNAL INEQUALITY IN HEIGHT.
(Epoch 1.6 days.)

Moon's declination.	Observed dh.	Computed dh.	Difference.
	Feet.	Feet.	Feet.
0°	0.0	0.0	0.0
12	1.4	1.4	0.0
21	2.3	2.2	0.1
25	2.5	2.5	0.0
22	2.3	2.3	0.0
13	1.7	1.5	0.2
0	0.0	0.0	0.0

The diurnal inequality in time I have tried to exhibit by numbers as well as by diagrams; it seems, however, that the incidental irregularities in the observations themselves, coupled with the fact that the observations generally were only made half-hourly and at other times hourly—so far exceed in magnitude the inequality itself as to make the effect of the changes of the moon's declination exceedingly obscure. The lunitidal intervals (for high and low water) between Oct. 17 and Dec. 28, 1853, between Jan. 28 and March 7, 1854, and between June 1 and July 7, 1854, were tabulated in vertical columns; the means of the alternate values were tabulated in the 2d column, and placed in the horizontal line opposite the intermediate value of column one. The numbers in the first column were next subtracted from the corresponding numbers in the second column, if the interval belonged to the inferior transit; if belonging to the superior, the values in the second column were subtracted from those in the first. The moon's declination, for noon each day, was also set down. The 276 values for diurnal inequality in time, thus obtained, were plotted. After attempting to deduce an epoch and arranging the values for different assumptions for epoch, no satisfactory result could be obtained in any way according with the expression

$$d\psi = \frac{g \tan. \delta'}{1 + A \cos.^2 \phi}$$ (see Lubbock, Phil. Trans. 1837),

and the results of the investigation must be confined to the following general remark. The diurnal inequality in time is in maxima probably not exceeding two hours; it seems to be less in amount for the times of high water than for the times of low water, a result the reverse of that belonging to the inequality in height. A similar conclusion was arrived at in the discussion of the tides at San Francisco, Cal. (Prof. A. D. Bache in Coast Survey Report for 1853, p. *81), when the *smaller* inequality in height of high water (when compared with that for low water) corresponded to the *greater* inequality in time of high water (when compared with the inequality for low water). Whether the inequality of the height for high or low water is the greater or smaller depends only on the epoch of the diurnal wave compared with the epoch of the semi-diurnal wave. There is no regular increase of the inequality corresponding to an increasing (irrespective of sign) declination of the moon, but the curve appears double-crested about the time of maximum declination, there being a sudden diminution in the inequality, preceded and followed by high values; about the time of the moon's crossing the equator the inequality is very irregular.

On Plate IV, the actual separation of the semi-diurnal and the diurnal wave has been effected graphically, for which purpose a part of the best observations was selected; these observations extend over the period from Oct. 30 to Nov. 22, 1853. The process of decomposition in use in the U. S. Coast Survey was at first an analytical one, by computing sine curves; since 1855, however, a graphical process, equivalent thereto, was substituted; this latter method, as introduced by assistant L. F. Pourtales, may be briefly explained as follows: After the observations were plotted and a tracing is taken, the traced curves are shifted in epoch 12 (lunar) hours forward, when a mean curve is pricked off between the observed and traced curves; this process is repeated after the tracing paper has been shifted 12 hours backward; the average or mean pricked curve thus obtained represents the semi-diurnal wave. On an axis parallel with that on which the time is counted, the differences between the originally observed and the constructed semi-diurnal wave were laid off; this constitutes the diurnal curve. In the case in hand I have simplified the process of separation by blackening the under surface of the tracing paper with a lead pencil, and running in with a free hand; the intermediate curves by the pressure of a style, an average of the two traces thus left on the lower paper, gave the semi-diurnal wave in quite an expeditious manner. On the diagram, the diurnal curve with its epoch of high water nearly coinciding with that of the semi-diurnal wave, appears plainly with its variation in size depending on the moon's declination.

Investigation of the Form of the Tide Wave.—The shape of the tide wave has been ascertained in the manner described in art. (479) Tides and Waves, and depends on the hourly observations of 60 tides, 30 during spring tides and an equal number during neap tides, that is, the observed heights on the day of the syzygies and quadratures and on the first and second day after, were tabulated, forming ten groups of three columns each, from low water to low water. The columns of an equal number of hours (they vary from 16 hours to 11 hours) were united in a mean. In order to combine these it was assumed that the interval from the observed low water to the next following low water corresponds to 360° of phase, and the time of every intermediate observation was converted into phase by that proportion. In order to render the observed heights comparable, the range from high to low water in every half tide (the reading of low water for phase 0 generally not being identical with the reading of the succeeding low water or phase 360°) was supposed to correspond to 2.00, and the elevation above the low water was converted into number by that proportion, thus furnishing a series of ordinates for equidistant abscissæ. The means of all the phases and corresponding converted depressions within every 30th degree of phase were then taken with proper regard to the weights, depending on the number of columns, of equal hours, united at the commencement of the reduction. By observation of the progress of the numbers,

RECORD AND REDUCTION OF THE TIDES.

it was easy to alter the latter so as to make them exactly correspond to the phases 30°, 60°, 90°, 120°, etc. In this manner the following numbers have been obtained:—

FOR THE SPRING-TIDE WAVE OCCURRING ONE AND A QUARTER DAY AFTER FULL AND NEW MOON.

Phase of groups.						Proportional height above low water.					
0⁵	0⁵	0⁵	0⁵	0⁵	Mean. 0⁵						Mean.
26	28	30	33	36	30	0.00	0.00	0.00	0.00	0.00	0.00
51	55	60	65	72	59	0.06	0.23	0.24	0.27	0.10	0.21
77	83	90	98	108	89	0.32	0.68	0.90	0.70	0.46	0.71
103	111	120	131	144	120	0.91	1.13	1.36	1.32	1.17	1.24
129	138	150	164	180	156	1.39	1.68	1.73	1.76	1.52	1.70
154	166	180	196	216	180	1.84	2.04	1.98	1.93	2.00	1.95
180	194	210	229	252	208	1.94	1.98	2.00	2.00	1.88	2.00
206	222	240	262	288	237	2.00	2.00	1.84	1.56	1.23	1.88
231	249	270	294	324	270	1.84	1.84	1.45	1.15	0.70	1.46
257	277	300	327	360	300	1.58	1.23	1.00	0.65	0.41	0.97
283	305	330	360		330	1.14	0.79	0.27	0.25	0.00	0.37
309	332	360			360	0.60	0.40	0.17	0.00		0.17
334	360					0.15	0.16	0.00			0.00
360						0.02	0.00				
						0.00					
Weight 5	4	13	7	1		Weight 5	4	13	7	1	

The columns headed "mean" show the ordinates of the waves for (nearly) equidistant intervals of time.

The following table contains the corresponding numbers for the neap tide wave occurring 1¼ day after the first and last quarter, and as derived from 30 tides observed hourly from low to low water:—

Phase of groups.						Proportional height above low water.					
0⁵	0⁵	0⁵	0⁵	0⁵	Mean.						Mean.
24	26					0.00					
48	51	28				0.07	0.00				
72	77	55	30	0⁵	29	0.26	0.17	0.00			
96	103	83	60	33	58	0.49	0.55	0.24	0.00		0.00
120	129	111	90	65	89	0.89	1.11	0.59	0.20	0.00	0.20
144	154	138	120	98	119	1.19	1.53	0.93	0.50	0.60	0.52
168	180	166	150	131	147	1.55	1.85	1.35	1.05	1.02	1.08
192	206	194	180	164	180	0.82	2.00	1.73	1.60	1.44	1.51
216	231	222	210	196	213	1.92	1.62	1.96	2.00	1.06	1.82
240	257	249	240	229	241	1.64	1.43	1.72	2.00	2.00	1.97
264	283	277	270	262	271	1.25	1.69	1.32	1.84	1.43	1.84
288	309	305	300	294	301	0.94	0.67	0.80	1.37	0.79	1.08
312	336	332	330	327	331	0.51	0.24	0.32	0.79	0.15	0.56
336	360	360	360	360	360	0.23	0.05	0.04	0.42	0.00	0.22
360						0.00	0.00	0.00	0.00	0.00	0.00
Weight 3	4	8	13	1	1	Weight 3	4	8	13	1	1

The results are represented in the annexed diagram. The result for the neap tide curve has also been multiplied by $\frac{47}{75}$, the ratio of neap and spring tide range as found on a preceding page, and was increased by 0.5 to refer it to the same level.

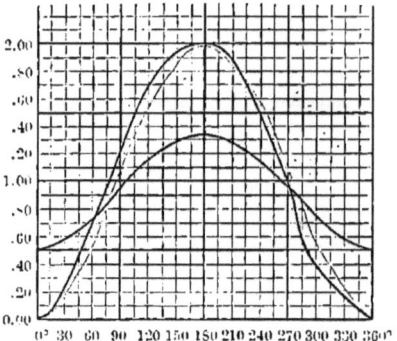

The full curves in the diagram show the form of the spring and neap tide wave (the scales being arbitrary), to which has been added for convenient comparison the dotted curve representing the neap tide wave on the same relative scale as the spring tide wave. It is apparent that the spring tide wave is slightly steeper between low and high water than between high and low water, and that the neap tide wave is very nearly symmetrical in respect to rise and fall.

We have seen that the duration of rise is $6^h\ 04^m.7$, hence the duration of fall will be $6^h\ 19^m.7$; or in making ebb the time is 15 minutes greater than in making flood, a circumstance in conformity with the shape of the curves of rise and fall. This holds good for an average tide; according to art. (510) Tides and Waves, if the place of observation is not far from the sea, or, as in our case, in a bay, the water will occupy a shorter time to rise than to fall, and the inequality will be greater at spring tides than at neap tides; this is fully illustrated in the preceding diagram, the spring tide wave being the steeper of the two.

The form of the tide waves will be found closely represented by the following expressions:—

For the spring tide wave—
$$5.83 + 5.58\ sin.(\theta + 278°) + 0.20\ sin.(2\theta + 281°);$$
For the neap tide wave—
$$2.42 + 2.25\ sin.(\theta + 269°) + 0.09\ sin.(2\theta + 290°);$$

in which expressions the angle θ counts from low water to low water, from 0 to 360°, and the height of the wave is expressed in feet.

The relative numbers, given above, as the ordinates, have been changed in the proportion of 2 to 11.1 for the higher and of 2 to 4.7 for the lower wave. The following table shows the agreement between observation and the numerical expressions, in which the 3d and higher terms are zero:—

FORM OF THE TIDE WAVE AT VAN RENSSELAER HARBOR.

Phase.	Height of Spring tide, in feet.		Height of neap tide, in feet.	
	Observed.	Computed.	Observed.	Computed.
0	0.0	0.1	0.0	0.1
30	1.2	1.4	0.5	0.4
60	3.9	3.9	1.3	1.3
90	6.9	6.8	2.5	2.5
120	9.4	9.3	3.5	3.5
150	10.7	10.9	4.3	4.3
180	11.1	11.1	4.6	4.6
210	10.4	10.2	4.3	4.4
240	7.9	8.0	3.7	3.7
270	5.4	5.3	2.5	2.5
300	2.1	2.4	1.3	1.5
330	0.9	0.5	0.5	0.4
360	0.0	0.1	0.0	0.1

Respecting the effect of the wind and ice on the tides, it may be remarked that the former can only be slight, since the sea is protected from the direct action of the wind by its icy cover for the greater part of the year. When the sea is partially open, the effect becomes sensible, as may be seen by the following note extracted from the log-book:—

"August 17, 1853. The above records show a heavy gale from the southward gradually hauling to the eastward; the effect of this gale on the tides was very marked; our flood rose two feet above any previous register, overflowing the ground ice, and our last ebb or outgoing tide was hardly perceptible." The ice crust cannot sensibly affect (by friction on its lower surface) the progress of the tide wave, and will certainly not sensibly interfere (by friction on the ice foot and breakage of the ice fields) with the rise and fall of the tide.

Progress of the Tide Wave.—The tide at Van Rensselaer Harbor may be taken as a derived tide, and transmitted to it from the Atlantic Ocean, and in part modified by the small tide originating in the waters of Baffin's Bay; which latter tide, however, must necessarily be small, particularly on account of the general direction of the bay, which is very unfavorable for the production of a tide wave. That the tide wave is travelling up along the western coast of Greenland, or, in other words, reaches Van Rensselaer Harbor from the southward, may be seen from the following observed establishments:—

Holsteinborg Harbor, latitude 66° 56', longitude 53° 42'. High water at F. & C. $6^h\ 30^m$. Spring tides rise 10 feet.—Capt. Inglefield, 1853.

Whalefish Islands (near Disco), latitude 68° 59', longitude 53° 13'. Time of high water F. & C. $8^h\ 15^m$. Highest tide 7½ feet.—Parry's 3d Voyage of Discovery.

Godhavn (Disco), latitude 69° 12', longitude 53° 28'. Tidal hour 9^h. Rise and fall 7½ feet.—See Map in Narrative of Kane's First Voyage.

Upernavik, latitude 72° 47', longitude 56° 03'. High water at F. & C. 11^h. Rise 8 feet.—Capt. Inglefield, 1854.

Wolstenholm Sound, latitude 76° 33', longitude 68° 56'. High water at F. & C. $11^h\ 8^m$. Rise, both at spring and neaps, 7 to 7½ feet.—(See Admiralty Chart of Baffin's Bay, sheet 1, 1853, corrected to 1859.) The observations themselves, taken by Captain Saunders of H. M. S. North Star, in 1849 and 1850, were kindly furnished to Prof. Bache by the Hydrographer to the Admiralty, Captain J. Washington, R. N., and are given in the appendix to this paper. And finally,

Van Rensselaer Harbor, latitude 78° 37′, longitude 70° 53′. High water at F. & C. $11^h\ 50^m$, as derived from the preceding analytical expression. Rise and fall at spring tide 11.1 feet, at neap tide 4.7 feet, average range 7.9 feet.

By means of the difference in the establishments of Holsteinborg and Van Rensselaer, we can obtain an approximation to the depth of Baffin's Bay and Smith's Straits, viz:—

	Tidal hour.	Longitude.	Sum.	Difference.	
Holsteinborg	$6^h\ 30^m$	$3^h\ 35^m$	$10^h\ 05^m$	$6^h\ 28^m$	Difference corrected for the moon's motion $6^h\ 26^m$.
Van Rensselaer	11 50	4 43	16 33		

Assuming the distance along the channel to be 770 nautical miles, we have a velocity of the tide wave of about 202 feet in a second, which, according to Airy's table (174), Tides and Waves, would correspond to a depth of nearly 1300 feet, or about 220 fathoms—a result probably smaller than the true value, since the other observations indicate a greater depth, it may be taken as an inferior limit; in the same manner we find from the co tidal hours of Upernavik and Van Rensselaer a depth of near 800 fathoms, and a similar result from the Wolstenholm observations; this last result may perhaps be taken as an upper limit.

Soundings.—The following soundings have been copied from the log-book:—

June 19, 1853. Lat. 51° 12′, long. 52° 8′ (government sounding twine and 32-pound shot).

Chronometer time.	Mark.
$8^h\ 47^m\ 0^s$	Red, started.
49 10	White.
52 10	Bottom, with 178 fathoms; shot brought up with gray mud and fine sand. The line was afterwards measured.

June 26, 1853. Lat. 59° 48′, long. 50° 3′ (government sounding twine and 32-pound shot).

Chronometer time.	Mark.	Chronometer time.	Mark.
$3^h\ 56^m\ 35^s$	Started 75 fathoms from the next mark.	$4^h\ 21^m\ 25^s$	White.
57 25	Red.	25 10	Red.
58 50	White.	29 15	White.
4 00 37	Red.	33 25	Red.
2 48	Black.	37 30	Black.
5 16	White.	42 0	White.
8 0	Red.	46 30	Red.
11 0	White.	51 15	White.
14 5	Red.	56 0	Red.
- - - - -	Missed the mark.	58 0	Bottom with 1817 fathoms, line cut.

August 1, 1858. Melville Bay, lat. 75° 40′, long. 62° 12′ (government sounding twine and 32-pound shot).

Chronometer time.	Mark.
$5^h\ 17^m\ 6^s$	Started.
18 8	Red.
49 40	White.
51 10	Red.
54 0	Black.
54 15	Bottom with 429 fathoms; shot brought up with dark green sand (specimen preserved).

www.ingramcontent.com/pod-product-compliance
Lightning Source LLC
Chambersburg PA
CBHW020307090426
42735CB00009B/1246